Dynamic Programming
(with management applications)

N. A. J. HASTINGS Ph.D., M.A., C.Eng., M.I.Mech.E.

Department of Engineering Production,
University of Birmingham

CRANE, RUSSAK & COMPANY, INC.
NEW YORK

519.7
H358

Published in the United States by
Crane Russak & Company, Inc.
52 Varderbilt Avenue
New York. N.Y. 10017
Library of Congress Catalog
Card No. 72 – 88945

First published in 1973

ISBN 0 – 8448 – 0116 – x

Printed in Hungary

Preface

Dynamic programming is concerned with the solution of optimisation problems which can be formulated as a sequence of decisions. Applications arise in such areas as production planning, stock control, component and equipment maintenance and replacement, company and investment planning, allocation of resources, product assortment and process design and control. There are many instances of other applications and new ones are continually being found.

This book aims to fulfil the current need for a textbook of Dynamic Programming which:

1. starts from an elementary level
2. deals thoroughly with problem formulation
3. contains plenty of worked examples and exercises and
4. deals with stochastic problems in reasonable depth.

The material has been developed from a course in Dynamic and Markov Programming which I have given for some years at the Department of Engineering Production in the University of Birmingham, with additional research material on stochastic problems.

The title 'Dynamic Programming' stems from the work of Richard Bellman, published largely in Bellman (1957), (1962) and Bellman and Dreyfus (1962). Bellman was the first to appreciate the wide range of applicability of a computational procedure which we refer to here as the *value iteration algorithm*. At much the same time Ronald Howard published an excellent monograph (Howard (1960)) which deals with the *policy iteration algorithm* and its application to probabilistic sequential decision problems. At the same time and subsequently many other books and papers have appeared which have served to temper and to extend these key works and to place their results in a wider context.

In this book, the value iteration algorithm is introduced in Chapter 1. Experience shows that students who can readily grasp the mathematics of the algorithm may nevertheless have considerable difficulty in problem formulation. Emphasis is therefore placed on the use of a standard system of terms and symbols, on the use of networks to represent problems and on a standard formulation procedure. The rather dogmatic statement of a Principle of Optimality favoured by Bellman is replaced by a set of validity conditions which apply to algorithms of the value iteration type.

Chapter 2 is concerned with further deterministic applications of the value iteration algorithm. Several examples are worked through and are used to illustrate additional features of the underlying technique. Examples are drawn from the areas of production planning, equipment replacement, product assortment and resource allocation.

In Chapter 3 the value iteration algorithm is applied to probabilistic problems. The term 'Markov programming' is used to describe the general area of optimisation of systems which undergo Markov processes and generate returns. Markov decision problems which continue for a finite number of stages are introduced with the aid of a marketing example. A short cut method is described in the context of a television guessing game and a worked example of a stochastic inventory problem is given.

Chapter 4 deals with infinite stage Markov programming in discrete time. An introduction to infinite stage Markov processes is given, using the fundamental matrix which compactly summarises the transient behaviour of systems. This approach is well suited to computational work. The models described in this and the following chapter are applicable to a wide range of probabilistic decision problems including finite queuing situations. Algorithms for finding gain and bias optimal policies are described and illustrated by examples.

Chapter 5 presents an elementary introduction to continuous time probabilistic decision problems. Discussion of this topic has hitherto been almost entirely confined to research literature. Algorithms are described for the solution of undiscounted and discounted semi-Markov decision problems and worked examples are given.

Preface

Chapters 1 to 3 require little specific mathematical foreknowledge although some familiarity with mathematical terminology and methods is assumed. Chapters IV and V assume an elementary knowledge of linear algebra and probability theory.

I am grateful to my colleagues at the University of Birmingham, particularly Professor K. B. Haley, Mr. C. S. Edwards and Mr J. M. C. Mello for providing a stimulating and sympathetic environment with a nice balance of academic and applied interests, and to Professor Paul J. Schweitzer whose work and correspondence I have found particularly valuable. I would also like to express my thanks to my wife, Christine, without whose unfailing support this book may never have been written.

N. A. J. Hastings

Contents

Contents

CHAPTER 1

The Value Iteration Algorithm

1.1 SYSTEMS, STATES AND PROCESSES

In the analysis of many operational problems it is convenient to consider the idea of a system which has a number of possible states and which makes a sequence of transitions between them. For example, in a problem of machine reliability and replacement the machine may be the system and a state may be defined by its age or condition of wear; in the analysis of a stock holding facility we may consider a system whose states correspond to the possible stock levels of a given item.

A state of a system may be defined in terms of one or more discrete or continuous variables. It is often either essential or adequate to consider discrete states, as for example where stock levels can only change by integer amounts. In the case of a machine which deteriorates gradually with time the state space is essentially continuous, but the application of a practical form of measurement imposes a discrete state discription. Thus in making a repair or replacement decision one might reasonably have a different rule for a four year old as opposed to a five year old machine, but a proliferation of rules based on age in weeks would be absurd.

The current work deals primarily with discrete state variables and with optimisation over finite sets. In this it differs from studies in engineering and natural science where continuous state variables are common and techniques which involve assumptions of differentiability are widely used. Again, in engineering and natural science it is often the case that a system which is in a given initial state and subject to known controls will follow a known, deterministic path. However, in operational problems one will often find that, even for a given initial state and environment, chance elements play a major part in determining the path followed by a system. We therefore divide our study between deterministic problems (Chapters 1 and 2) and stochastic problems (Chapters 3 to 5).

In general we consider a system with a finite set of discrete states which can follow any of a number of relevant paths or processes. The movement of the system is controlled, or at least influenced, by a decision maker who, at each stage, uses one of a set of feasible actions. As the system proceeds it generates a sequence of returns, which will usually be a distance, a time, an income or expenditure of money, the yield of a product or the consumption of a resource. The decision maker wishes to find the sequence of actions which in some sense optimises a function of the returns generated by the system.

The type of technique which can be brought to bear on such an optimisation problem depends on how many possible states the system has, how they are interconnected and on the optimisation criterion. Dynamic programming is concerned with problems where the states are not too numerous, the possible processes each form an ordered progression and the optimisation criterion is of a certain type, typically the sum or discounted sum of the returns.

1.2 KEY TERMS

We now introduce more formally the system of terminology and symbols which runs throughout this book. Use is made of a nominal problem of finding the shortest path through a simple network. The problem is as follows.

Figure 1.1 shows a network of cities represented by circles and roads represented by lines. The lengths of the roads are indicated (the figure is not to scale). A man at city A wishes to travel to city H by the shortest possible path. He must always move in a left to right direction as indicated by the arrows. What path should he take?

We wish not only to solve this problem but also to develop a method for solving all problems of a similar type. Our approach will be such as to lay the foundations of a system of terminology by which we can show that many problems have the same essential structure as the current one. Our first step is to introduce this terminology and to interpret it in terms of the shortest path problem.

State. A state is a configuration of a system and is identified by a label which indicates the features or properties corresponding to that state. In the shortest path problem a city is a state.

Stage. Dynamic programming is concerned with systems which undergo processes involving a series of moves from one state to another. A stage is a single step in such a process, and corres-

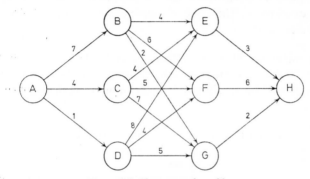

Figure 1.1. Shortest path problem

ponds to the transition of a system from one state to an adjacent state. In Figure 1.1 the movement of the traveller from one city to the next comprises one stage. Every path from A to H has three stages.

Action. At each state there is a set of actions available from amongst which a choice must be made. For example at state C in the shortest path problem there are three possible actions, namely, to go to E, F or G. Solving a dynamic programming problem involves finding the best sequence of actions in accordance with some given objective.

Plan. A plan is a set of actions one for each of a number of states. An optimal plan is the best set of actions in accordance with a given objective. Some authors use the term 'policy' rather than 'plan' but we reserve 'policy' for infinite horizon problems.

Return. A return is something which a system generates over one stage of a process. In the shortest path problem it is the distance covered in passing from one city to the next. A return is usually something like a profit, a cost, a distance, a yield of a product or a consumption of a resource, etc.

Value of a state. The value of a state is a function (often simply the sum) of the returns generated when the system starts in that state and a particular plan is followed. In the shortest path example the value of a state under a given plan corresponds to the distance from that city to the terminal city when that plan is followed. The value is given by the sum of the distances over the various stages on route. The value of a state under an optimal plan is the optimal value.

As we encounter different problems we shall relate them to the basic structure of dynamic programming by specifying them in the terms just described. The specification of the shortest path problem is in Table 1.1.

Table 1.1. SHORTEST PATH PROBLEM: SPECIFICATION

State	A city
Stage	A transition from a city to an adjacent city
Action	Taking a particular route from a city
Return	Distance from a city to an adjacent city
Value of a state	Distance from the corresponding city to the terminal city under a given plan

1.2.1 Network Representation

The network of Figure 1.1 is a convenient representation of the form of the shortest path problem. We shall often make use of networks, even where the structure is at first less obvious than in this case. In general, the states of a system can be represented by nodes in a network, with arcs representing the possible transitions and numbers against the arcs representing the returns. An initial (source) node has no arcs leading into it and a terminal (sink) node has no arcs leading from it.

The value iteration algorithm in its simplest form is concerned with problems which can be represented by serial networks. In a serial network the nodes can be grouped into disjoint sets $u(n)$, $n = 0, 1, \ldots, m$. $u(m)$ is the set of all initial nodes, $u(0)$ is the set of all terminal nodes and every arc from a node in set $u(n)$ leads into a node in set $u(n-1)$. In Figure 1.1 the sets of nodes are

$$
\left.
\begin{aligned}
u(0) &= \{H\} \\
u(1) &= \{E, F, G\} \\
u(2) &= \{B, C, D\} \\
u(3) &= \{A\}
\end{aligned}
\right\} \tag{1.1}
$$

Symbols

In Figure 1.1 we indicated the various states by letters A, B, C, etc. For computational work we use stage and state variables. The stage variable, n, indicates the number of stages which remain until a terminal state is reached, so that the states in set $u(n)$ all have the

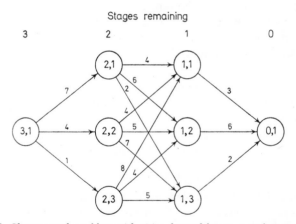

Figure 1.2. Shortest path problem with cities denoted by stage and state variables

same stage number, n. The state variable, i, is used to distinguish the individual states at each stage. In general the i th state at stage n is called state (n, i). The shortest path problem with this state numbering

is shown in Figure 1.2. The stagewise sets with numerical labelling
are as follows:

$$\left.\begin{array}{l} u(0) = \{(0, 1)\} \\ u(1) = \{(1, 1), (1, 2), (1, 3)\} \\ u(2) = \{(2, 1), (2, 2), (2, 3)\} \\ u(3) = \{(3, 1)\} \end{array}\right\} \tag{1.2}$$

To define a particular state we need the two numbers (n, i) but we may
talk about 'state i' in contexts where no ambiguity exists about the
stage number.

At a general state there is a set K of possible actions k. We may
sometimes wish to identify precisely the set of actions available at
a particular state (n, i) and we can call this set K_{ni} where

$$K_{ni} = \{1, 2, \ldots, k, \ldots, k_{ni}\} \tag{1.3}$$

However, the subscripts can normally be omitted without ambiguity
since the relevant state is implied by the context.

In the shortest path problem we number the roads leading from
each state in a clockwise sequence, starting from the 12 o'clock position. Action k consists in choosing road k. Thus for state $(2,1)$ actions
1, 2, 3 respectively result in transition to states $(1, 1)$, $(1, 2)$, $(1, 3)$
respectively. At state (n, i) the successor state $(n-1, j)$ under action
k is determined by the equation

$$j = k \tag{1.4}$$

In general the successor state is determined from the current stage,
state and action variables by a function called the transition function,
t. The equation relating the value of j to the stage, state and action
variables n, i, k is called the transition equation and has the form

$$j = t(n, i, k) \tag{1.5}$$

Equation 1.4 is a very simple transition equation.

When an action is chosen at a given state the return at the current
stage is determined. The return associated with state (n, i) and action
k is $r(n, i, k)$, where r is the return function.

The value of state (n, i) under an optimal plan is denoted by $f(n, i)$.
f is the optimal value function. The specification of the shortest path
problem showing the nomenclature just defined is given in Table 1.2.

Table 1.2. SPECIFICATION AND SYMBOLS FOR THE SHORTEST PATH PROBLEM

Stage	A transition from a city to an adjacent city	n
State	A city	(n, i)
Action	Taking a particular round from a city	k
Return	Distance from a city to an adjacent city	$r(n, i, k)$
Optimal value of a state	Distance from the corresponding city to the terminal city under an optimal plan	$f(n, i)$

1.3 SOLUTION OF THE SHORTEST PATH PROBLEM BY VALUE ITERATION

In broad outline the method of solution is as follows. We give the terminal state the value zero. We then determine the optimal action and optimal value for each state at stages 1, 2, etc, working backwards through the network until all the states have been dealt with. As we go we enter the optimal actions and values in a table. Finally we pick out the optimal path from the table by starting from the initial state and following the optimal plan just determined.

In detail the method is as follows. We give the terminal state the value zero, that is $f(0, 1) = 0$. In states $(1, 1)$, $(1, 2)$ and $(1, 3)$ there is only one possible action, $k = 1$, and the optimal values of these states are the direct distances to the terminal state, namely, $f(1, 1) = 3$, $f(1, 2) = 6$, $f(1, 3) = 2$. These results are shown in the top three rows of Table 1.3. Italics indicate that an action or value is optimal.

At stage 2 we consider each state in turn, starting with state $(2, 1)$. At state $(2, 1)$ there are three possible actions, $k = 1, 2, 3$, which result in transition to states $(1, 1)$, $(1, 2)$, $(1, 3)$ respectively. We consider them in turn. For action 1 the value of state $(2, 1)$ is given by the sum of the distance from $(2, 1)$ to $(1, 1)$ and the distance from $(1, 1)$ to the terminal state, that is

$$r(2, 1, 1) + f(1, 1) = 4 + 3 = 7 \tag{1.6}$$

Similarly for actions 2 and 3, the distances are

$$r(2, 1, 2)+f(1, 2) = 6+6 = 12 \tag{1.7}$$

$$r(2, 1, 3)+f(1, 3) = 2+2 = 4 \tag{1.8}$$

The aim is to find the shortest path to the terminal state. We therefore pick out the smallest of the three quantities shown in equations 1.6, 1.7 and 1.8. This is given by equation 1.8 and corresponds to action 3. The value of state (2, 1) under this action is the optimal value and is

$$f(2, 1) = 4 \tag{1.9}$$

The computation of this result is shown in Table 1.3 in rows corresponding to state (2, 1). The trial values of the test quantity

$$r(2, 1, k)+f(1, k) \tag{1.10}$$

under actions $k = 1, 2, 3$, appear in rows 3, 4 and 5 respectively. The optimal action and value are in italics. By giving state (2, 1) the optimal value, $f(2, 1) = 4$, we effectively discard actions 1 and 2 at state (2, 1) from all further consideration. This is valid in any problem involving the minimisation or maximisation of the sum of independent returns in a serial network. This point is discussed more fully later in this chapter, where general conditions governing the use of the value iteration method are given. If there is a tie between two actions we may choose either arbitrarily, but we make it a convention to retain the first optimal action found unless some other rule for resolving ties is stated.

We can summarise the method of obtaining the optimal action and value for state (2, 1) by the equations

$$f(2, 1) = \min_{k \in K} [r(2, 1, k)+f(1, k)] \tag{1.11}$$

$$K = \{1, 2, 3\} \tag{1.12}$$

Equation 1.11 means that we choose the action k which makes the expression in the square brackets a minimum. It is called a recurrence relation because it, and its generalisation to state (n, i), are used recursively to calculate optimal state values at stage n from the optimal state values at stage $n-1$.

Continuing with the shortest path problem, at state (2, 2) the trial values under actions $k = 1, 2, 3$, are

$$\left.\begin{array}{l} r(2, 2, 1)+f(1, 1) = 4+3 = 7 \\ r(2, 2, 2)+f(1, 2) = 5+6 = 11 \\ r(2, 2, 3)+f(1, 3) = 7+2 = 9 \end{array}\right\} \qquad (1.13)$$

Action 1 is optimal and $f(2, 2) = 7$. Similarly for state (2, 3) the actions $k = 1, 2, 3$, respectively give

$$\left.\begin{array}{l} r(2, 3, 1)+f(1, 1) = 11 \\ r(2, 3, 2)+f(1, 2) = 10 \\ r(2, 3, 3)+f(1, 3) = 7 \end{array}\right\} \qquad (1.14)$$

Action 3 is optimal and $f(2, 3) = 7$. These calculations are shown in Table 1.3. rows 7 to 9 and 10 to 12.

Table 1.3. SHORTEST PATH PROBLEM: FULL TABLE OF CALCULATIONS

Stage	State	Action	Trial Value
1	1	*1*	*3*
1	2	*1*	*6*
1	3	*1*	*2* ✓
2	1	1	4+3 = 7
2	1	2	6+6 = 12
2	1 ✓	3	2+2 = 4 ✓
2	2	*1*	4+3 = 7
2	2	2	5+6 = 11
2	2	3	7+2 = 9
2	3	1	8+3 = 11
2	3	2	4+6 = 10
2	3	3	5+2 = 7
3	1	1	7+4 = 11
3	1	2	4+7 = 11
3	1	3	1+7 = 8

2*

The calculation of optimal actions and values continues in a similar way over the remainder of the network. At a general state (n, i) the procedure is summarised by the recurrence relation,

$$f(n, i) = \min_{k \in K} [r(n, i, k) + f(n-1, k)] \tag{1.15}$$

In the present problem only state $(3, 1)$ remains to be considered. There, actions $k = 1, 2, 3$ give the following trial values

$$\left. \begin{array}{l} r(3, 1, 1) + f(2, 1) = 7 + 4 = 11 \\ r(3, 1, 2) + f(2, 2) = 4 + 7 = 11 \\ r(3, 1, 3) + f(2, 3) = 1 + 7 = 8 \end{array} \right\} \tag{1.16}$$

Action 3 is optimal and $f(3, 1) = 8$. These calculations complete Table 1.3.

The optimal path is picked out by entering Table 1.3 at the initial state $(3, 1)$ and noting the optimal action there, namely action 3; this tells us that the next state on the optimal path is state $(2, 3)$. We then go to state $(2, 3)$ in the table where we find that action 3 is optimal so the next state is $(1, 3)$; at state $(1, 3)$ the optimal (and only) action is action 1 which takes us to the terminal state. The optimal path is $(3, 1)$ to $(2, 3)$ to $(1, 3)$ to $(0, 1)$, and is listed in Table 1.4. Table 1.3 enables us to find the optimal path from any city to the terminal city and the length of that path. Such a set of paths is called a minimal spanning tree.

Table 1.4. SHORTEST PATH PROBLEM; OPTIMAL PATH

Stage	State	Action	Value
3	1	3	8
2	3	3	7
1	3	1	2

1.4 THE VALUE ITERATION METHOD

The method of solution just described for the shortest path problem is the value iteration method. It is the main computational algorithm of dynamic programming. It is summarised by the following algebraic statements.

Recurrence relation:

$$f(n, i) = \min_{k \in K} [r(n, i, k) + f(n-1, j)] \qquad (1.17)$$

Transition equation:

$$j = t(n, i, k) \qquad (1.18)$$

Terminal values:

$f(0, i)$ given for all terminal states.

Any finite, serial shortest path problem can be solved by repeated application of equation 1.17. In words the recurrence relation is

$$\begin{matrix} \text{Value of} \\ \text{a state} \end{matrix} = \begin{matrix} \text{Minimum over all} \\ \text{relevant actions} \end{matrix} \left[\text{Return} + \begin{matrix} \text{Value of} \\ \text{next state} \end{matrix} \right]$$

The value iteration method applies generally, though with variations in detail, to sequential problems over a wide range of applications. It may sometimes be convenient to combine equations 1.17 and 1.18 and express the recurrence relation as

$$f(n, i) = \min_{k \in K} [r(n, i, k) + f(n-1, t(n, i, k))] \qquad (1.19)$$

A flow chart for the algorithm is shown in Figure 1.3. The flow chart has three loops. The inner loop (loop 3) corresponds to finding the best action at a state. The middle loop (loop 2) takes the procedure from one state to another at each stage, whilst the outer loop (loop 1) takes the procedure over each stage in turn. By following the flow chart in Figure 1.3 we determine the optimal action and value for every state, information corresponding to the italic numbers in Table 1.3 in the shortest path problem. The optimal path or process from a given initial state can be picked out in the manner already described for Table 1.4. The general procedure for picking out the optimal process is summarised by the flow chart in Figure 1.4.

12

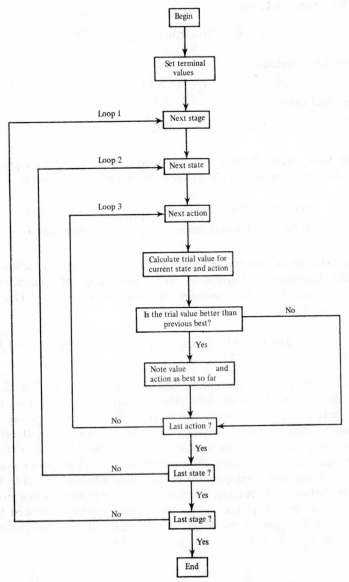

Figure 1.3. Flow chart for the value iteration algorithm

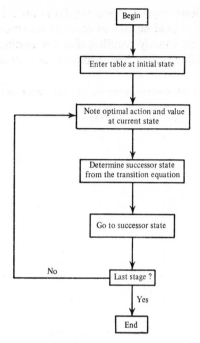

Figure 1.4. Determination of the optimal process from the full table of optimal actions and values

1.5 MAXIMISATION

The value iteration method can be used for maximisation as readily as for minimisation. In maximisation problems the action which yields the largest trial value is found at each state and that value and action are optimal. The recurrence relation is

$$f(n,i) = \underset{k \in K}{\text{Max}} [r(n,i,k)+f(n-1,j)] \qquad (1.20)$$

where in general

$$j = t(n, i, k)$$

As an example of a maximisation problem, suppose that in Figure 1.2 the traveller is a carrier and the returns represent the amounts he can earn at the various stages of a journey from the initial to the terminal city. What path should he take in order to maximise earnings?

The calculation exactly parallels that for finding the shortest path but the maximising actions and values are now optimal. The full table

Table 1.5. MAXIMISATION PROBLEM: FULL TABLE OF CALCULATIONS

Stage	State	Action	Trial Value
1	1	1	3
1	2	1	6
1	3	1	2
2	1	1	4+3 = 7
2	1	2	6+6 = 12
2	1	3	2+2 = 4
2	2	1	4+3 = 7
2	2	2	5+6 = 11
2	2	3	7+2 = 9
2	3	1	8+3 = 11
2	3	2	4+6 = 10
2	3	3	5+2 = 7
3	1	1	7+12 = 19
3	1	2	4+11 = 15
3	1	3	1+11 = 12

of calculations is shown in Table 1.5. The optimal path is (3, 1) to (2, 1) to (1, 2) to (0, 1) and the total return is 19 units. This result is shown in Table 1.6.

Table 1.6. MAXIMISATION PROBLEM: OPTIMAL PATH

Stage	State	Action	Value
3	1	1	19
2	1	2	12
1	2	1	6

1.6 DISCOUNTED RETURNS

In management decision problems it is frequently the practice to discount returns which will be received in a year or more's time. Suppose that £1 invested for one year yields interest £r. r is the interest rate. A sum of £$(1+r)$ to be received in one year's time can be regarded as having the same *present value* as a sum of £1 received now. In general the present value of £x received in one year's time is £$x/(1+r)$. Define the *discount factor*, b, by

$$b = 1/(1+r) \tag{1.21}$$

Assume that

$$r > 0$$

and

$$0 \leqslant b < 1 \tag{1.22}$$

If interest is compounded annually the present value of £x received after n years is £$b^n x$.

The value iteration method can be used to maximise or minimise the sum of discounted returns. Consider the following example.

1.6.1 Capital investment problem

A company has a plant which at present has capacity level 1. It is planning the future of the plant over a three year period. In the first year the company has the choice of the following actions,

1. Maintain capacity at level 1 and have a net return of 2 units,
2. Expand capacity to level 2 and have a net return (allowing for the cost of expansion) of -7 units.

Similar decisions can be made in subsequent years. The choice of actions and the associated returns are shown in Figure 1.5. The maximum capacity is level 3 and the possibility of reducing capacity at any point in the planning period is not considered; these assumptions are made solely to reduce the amount of arithmetic in the example. The value of the plant at the end of the planning period is assessed relative to its value at capacity level 1, and is 4 units for level 2 and 8 units for level 3. Note that some assessment of terminal values must be made

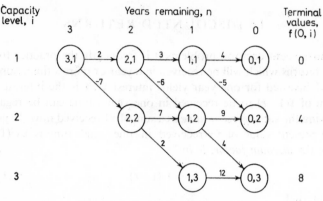

Figure 1.5. Capital investment problem

in problems of this type. However, the solution will not usually be very sensitive to small changes in the terminal values in discounted problems. Determine the investment programme which has the maximal present value, assuming an interest rate of $33\frac{1}{3}\%$.

Solution

The specification of the problem in standard terms and symbols is shown in Table 1.7. The recurrence relation is

$$f(n, i) = \underset{k \in K}{\text{Max}} [r(n, i, k) + bf(n-1, j)] \qquad (1.23)$$

Table 1.7. CAPITAL INVESTMENT PROBLEM: SPECIFICATION

Stage	A year. n = number of years remaining until the planning horizon is reached	n
State	Plant capacity level, i, when n years remain	(n, i)
Action	1 = maintain current capacity level 2 = expand to next higher capacity level	k
Return	Nett earnings in a year	$r(n, i, k)$
Optimal value of a state	Total discounted return when the current state is the starting state and an optimal plan is followed	$f(n, i)$

where b is the discount factor

$$b = 1/(1+r) = 0\cdot75 \tag{1.24}$$

The transition equation is

$$j = i+k-1 \tag{1.25}$$

The action set is $K = \{1, 2\}$ except in state $(1, 3)$ where $K = \{1\}$. The terminal values are

$$\left. \begin{array}{l} f(0, 1) = 0 \\ f(0, 2) = 4 \\ f(0, 3) = 8 \end{array} \right\} \tag{1.26}$$

The calculations follow the usual value interation scheme shown in the flow chart, Figure 1.3. As an example of the determination of the optimal actions and values consider state $(1, 2)$. The trial values under actions 1 and 2 respectively are

$$\left. \begin{array}{l} r(1, 2, 1)+0\cdot75\times v(0, 2) = 9+0\cdot75\times4 = 12 \\ r(1, 2, 2)+0\cdot75\times v(0, 3) = 4+0\cdot75\times8 = 10 \end{array} \right\} \tag{1.27}$$

Action 1 is optimal and $f(1, 2) = 12$. The full set of calculations is shown in Table 1.8, where the optimal actions and values are *italicised*. The optimal process is picked out from Table 1.8 in the manner shown in Figure 1.4. The optimal process is given in Table 1.9 and involves no expansion of capacity.

1.6.2 Discounted Cash Flow

Capital investment problems are sometimes analysed on the basis of the discounted cash flow (d.c.f.) rate of return. This is defined as follows. Consider a sequence of annual returns $r(1)$, $r(2)$, ..., $r(n)$. Their present value, f, at discount factor b is

$$f = r(1)+br(2)+b^2r(3)+\ldots+b^{n-1}r(n) \tag{1.28}$$

The d.c.f. rate of return (internal rate of return) is the interest rate which reduces f to zero. In general there may be no value or several values of the discount factor (and hence the interest rate) which make f zero. However, the concept is commonly applied to the case where

Table 1.8. CAPITAL INVESTMENT PROBLEM: CALCULATIONS

Stage	State	Action	Trial Value	
1	1	*1*	$4+0\cdot75\times0$	$= 4$
1	1	2	$-5+0\cdot75\times4$	$= -2$
1	2	*1*	$9+0\cdot75\times4$	$= 12$
1	2	2	$4+0\cdot75\times8$	$= 10$
1	3	*1*	$12+0\cdot75\times8$	$= 18$
2	1	*1*	$3+0\cdot75\times4$	$= 6$
2	1	2	$-6+0\cdot75\times12$	$= 3$
2	2	*1*	$7+0\cdot75\times12$	$= 16$
2	2	2	$2+0\cdot75\times18$	$= 15\cdot5$
3	1	*1*	$2+0\cdot75\times6$	$= 6\cdot5$
3	1	2	$-7+0\cdot75\times16$	$= 5$

$r(1)$ is negative and $r(2)$ to $r(n)$ are positive, in which case the d.c.f. rate of return exists and is unique. It can be found by numerical analysis.

Table 1.9. CAPITAL INVESTMENT
PROBLEM: OPTIMAL PROCESS

Stage	State	Action	Value
3	1	1	6·5
2	1	1	6
1	1	1	4

1.7 FORWARD RECURRENCE

The value iteration method in the form so far described involves starting from one or more terminal states of known value and working through the problem backwards. This is known as backward recurrence. Deterministic problems can equally well be solved by a system of forward recurrence. To illustrate the procedure consider again the shortest path example of Figure 1.1. In forward recurrence the stages are number forwards. The stage variable, n, indicates the number of stages from an initial state. As before, the state variable, i, indicates the state number within a given stage. The shortest path problem with this state numbering is shown in Figure 1.6.

An action at state (n, i) now corresponds to approaching that state along a given road. Let the roads into the states in Figure 1.6 be num-

Stages completed

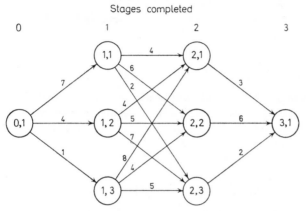

Figure 1.6. Shortest path problem numbered for forward recurrence

bered anticlockwise from the 12 o'clock position. Action k corresponds to approaching a state along road k. The transition equation relates the predecessor state $(n-1, j)$ to the current state (n, i) under action k. Its general form is

$$j = t(n, i, k)$$

and in the current example it is,

$$j = k$$

As before the return associated with state (n, i) and action k is denoted by $r(n, i, k)$, but it is now the distance to the current city from the previous city. The optimal value of state (n, i), denoted as before by $f(n, i)$, is now defined as the length of the shortest path from the initial to the current state. The specification of the optimal path problem in a forward formulation is shown in Table 1.10. The forward recurrence relation is

$$f(n, i) = \operatorname*{Min}_{k \in K} [r(n, i, k) + f(n-1, k)] \tag{1.29}$$

Table 1.10. SHORTEST PATH PROBLEM: SPECIFICATION
OF THE FORWARD FORMULATION

Stage	A step from a city to an adjacent city n = number of stages from initial city	n
State	A city	(n, i)
Action	Arriving at a city by road k	k
Return	Distance to current city from previous city	$r(n, i, k)$
Optimal value of a state	Distance to the corresponding city from the initial city under an optimal plan	$f(n, i)$

This equation is identical in appearance to the backward recurrence relation, equation 1.15, but as the symbols have been redefined the interpretation of the two equations is not the same. For example, the distance from city F to city H, Figure 1.1, is denoted by $r(1, 2, 1)$ in the backward formulation and by $r(3, 1, 2)$ in the forward formulation. The full calculations in the solution of the optimal path problem by forward recurrence are shown in Table 1.11.

Table 1.11. OPTIMAL PATH PROBLEM:
FORWARD RECURRENCE CALCULATIONS

Stage	State	Action	Trial Value
1	1	*1*	7
1	2	*1*	4
1	3	*1*	1
2	1	1	$4+7 = 11$
2	1	2	$4+4 = 8$
2	1	3	$1+8 = 9$
2	2	1	$7+6 = 13$
2	2	2	$4+5 = 9$
2	2	*3*	$1+4 = 5$
2	3	1	$7+2 = 9$
2	3	2	$4+7 = 11$
2	3	*3*	$1+5 = 6$
3	1	1	$3+8 = 11$
3	1	2	$6+5 = 11$
3	1	3	$2+6 = 8$

In practice, forward recurrence is advantageous when a deterministic problem has to be solved several times with different planning horizons. This may occur because a plan is periodically reviewed or where the appropriate horizon is unknown and a sensitivity analysis is undertaken. The value table can be extended forward in time without earlier calculations having to be repeated. An illustration is D. J. White and J. M. Norman: An example of problem embedding in deterministic dynamic programming. Opl. Res. Q. 20, 469–476 (1969). Where there is no special reason for choosing either formulation the backward formulation is normally used.

1.8 PROGRESSIVE PROBLEMS

So far we have dealt with serial problems where from every state a terminal state is reached in some predetermined number of stages, denoted in general by n. The value iteration method extends readily to *progressive* problems in which from each state a terminal state is

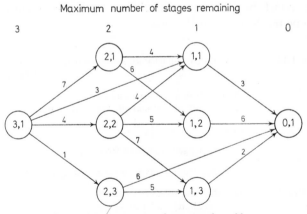

Figure 1.7. Progressive shortest path problem

reached in *not more than* a predetermined number of stages. From a state from which a terminal state is reached in n stages or less, transitions can only be made to states from which the terminal state is reached in w stages or less, where $w < n$. The shortest path problem shown is Figure 1.7 is progressive. The states can be arranged in columns and numbered in accordance with the maximum number of remaining stages. This gives a diagram similar to that of Figure 1.2 but with some stages involving moves to the right by more than one column.

The stage variable n now denotes the maximum number of stages remaining. The states can be divided into sets $u(n)$, $n = 0, 1, \ldots, m$. The members of $u(0)$ are terminal states. Any transition from a state in set $u(n)$, $n > 0$, takes the system to a state in set $u(w)$ where $w < n$. State (n, i) is the ith state in set $u(n)$. The transition equation has the form

$$(w, j) = t(n, i, k) \tag{1.30}$$

The successor state (w, j) is determined from the current state and action by the transition function, t, which may, of course, be expressed as a look up table or diagram. In the progressive shortest path problem an action k corresponds to leaving a state by road k, the roads being number clockwise from the 12 o'clock position. Successor states can be determined by reference to Figure 1.7. The formulation is specified in Table 1.12. A state is said to be adjacent to a given state if it can be reached in a single stage.

Table 1.12. SPECIFICATION OF THE PROGRESSIVE SHORTEST PATH PROBLEM

Stage	A step from a city to an adjacent city n = maximum number of stages remaining	n
State	A city	(n, i)
Action	Taking a particular road from a city	k
Return	Distance from a city to an adjacent city	$r(n, i, k)$
Optimal value of a state	Distance from the corresponding city to a terminal city under an optimal plan	$f(n, i)$

We wish to determine the shortest path from state $(3, 1)$ to state $(0,1)$. The recurrence relation is

$$f(n, i) = \underset{k \in K}{\text{Min}} \, [r(n, i, k) + f(w, j)] \tag{1.31}$$

Solution by the value iteration method follows similar lines to the serial case. The full calculations are shown in Table 1.13, and the optimal path is shown in Table 1.14.

Table 1.13. PROGRESSIVE SHORTEST PATH PROBLEM: CALCULATIONS

Stage	State	Action	Trial Value
0	1	—	0
1	1	1	3
1	2	1	6
1	3	1	2
2	1	1	4+3 = 7
2	1	2	6+2 = 12
2	2	1	4+3 = 7
2	2	2	5+6 = 11
2	2	3	7+2 = 9
2	3	1	6+0 = 6
2	3	2	5+2 = 7
3	1	1	7+7 = 14
3	1	2	3+3 = 6
3	1	3	4+7 = 11
3	1	4	1+6 = 7

Table 1.14. PROGRESSIVE SHORTEST PATH PROBLEM: OPTIMAL PATH

Stage	State	Action	Value
3	1	2	6
1	1	1	3

As a sample calculation consider state (2, 3). The values of states (0, 1) and (1, 3) are

$$f(0, 1) = 0 \atop f(1, 3) = 2 \Big\} \qquad (1.32)$$

The trial values for state (2, 3) under actions $k = 1$ and $k = 2$ respectively are

$$\left.\begin{array}{l} r(2, 3, 1)+f(0, 1) = 6+0 = 6 \\ r(2, 3, 2)+f(1, 3) = 5+2 = 7 \end{array}\right\} \tag{1.33}$$

Action 1 is optimal and $f(2, 3) = 6$.

In some progressive problems it is convenient to number the states $i = 1, 2, \ldots, m$. An example is the set partitioning problem in Chapter 2.

1.9 COMPUTATIONAL LOAD

Consider a serial shortest path problem with n stages, N states per stage and k actions per state. In solving such a problem by value iteration we make k additions and k comparisons at each of nN states. The total number of steps required, counting each addition and each comparison as one step is

$$2nNk \tag{1.34}$$

Such a problem could be solved by full enumeration, that is by calculating the lengths of all possible paths and picking the shortest. The number of steps required for this is as follows. There are Nk^n paths and to calculate the length of each requires n additions. We must then make one comparison per path, hence the total number of steps is

$$(n+1)Nk^n \tag{1.35}$$

In all but the smallest problems the savings of the value interation method are very significant. For example, if $n = N = k = 10$, expression 1.35 is of the order of 10^{12} and expression 1.34 is 2000. The savings achieved by dynamic programming arise from the fact that as the iteration procedes more and more suboptimal plans are identified and then disregarded for the remainder of the calculation.

Although dynamic programming is quicker than full enumeration it is itself limited in the size of problem it can handle. For example, if $n = N = k = 1000$ the number of computational steps required is 2×10^9. Such a problem would take over 5 hours to solve at one step per 10 microseconds.

Although the computation of a trial value can involve as little as a single addition, it will often in practice involve further steps to calculate the return and transition functions. On the other hand, the case $n = N = k = 1000$ involves a full network, whereas networks are often quite sparse.

1.9.1 Storage

In the value iteration method the optimal action and value of every state is stored. The amount of storage required is twice the number of states. For a problem of the type just discussed $2nN$ storage locations are needed. This requirement relates to calculations only and additional storage must be available for data and the programme, but these latter requirements are usually modest. Only the state values at the current and successor stages need be kept in fast access store.

1.10 VALIDITY CONDITIONS

(Exercises 1.1 to 1.6 may be undertaken before reading the rest of this chapter).

The *value of a state* has been defined as a function of the returns generated by a system when that state is the starting state and a given plan is followed. In the problems considered so far the returns have been independent and additive with or without discounting. Thus the distance from state (n, i) to state $(n-1, j)$ in a shortest path problem does not depend on any other variable, and the length of any path is equal to the sum of the relevant stage lengths. With problems of this type the validity of the value iteration method is easy to establish. The limitations on the validity of the method are less obvious, however, and we therefore consider general conditions which must be satisfied if the value iteration method is to be applicable. In the algebraic work which follows it is convenient to denote the state variable at stage n by i_n and an action at state (n, i_n) by k_n.

Consider a system undergoing a process in which at a general stage the system goes from state (n, i_n) to state $(n-1, i_{n-1})$ under action k_n, and generates a return $r(n, i_n, k_n)$. Let A_n be a general plan which determines a sequence of actions, $k_n, k_{n-1}, \ldots, k_1$. Suppose that the overall process has m stages and that we wish to maximise a function ϕ_m of the stage returns, i.e. to find the maximal value $f(m, i_m)$, defined by

$$f(m, i_m) = \max_{A_m \in W_m} [\phi_m(r(m, i_m, k_m), \ldots, r(n, i_n, k_n), \ldots, r(1, i_1, k_1))]$$

(1.36)

3*

where W_m is the set of all plans which start in state (m, i_m). In order that the maximum required in equation 1.36 can be found by value iteration two conditions must hold, namely the Separability Condition and the Optimality Condition.

1.10.1 Separability Condition

The separability condition governs whether or not the value of a state can be calculated by a recursive algorithm, given a fixed plan. Its general statement is;

For every plan the value of every state must be expressible as a function of the immediate return and the value of the succeeding state.

Let the value of state (m, i_m) under plan A_m be denoted by $f(m, i_m, A_m)$. Then

$$f(m, i_m, A_m) = \phi_m(r(m, i_m, mk_m), \ldots, r(n, i_n, k_n), \ldots, r(1, i_1, k_1)) \tag{1.37}$$

If the separability condition holds then for every state (n, i_n) and plan A_n we can write equation 1.37 in the form

$$f(n, i_n, A_n) = \Phi_n(r(n, i_n, k_n), f(n-1, i_{n-1}, A_{n-1})) \tag{1.38}$$

where

$$f(n-1, i_{n-1}, A_{n-1}) = \phi_{n-1}(r(n-1, i_{n-1}, k_{n-1}), \ldots r(1, i_1, k_1)) \tag{1.39}$$

and Φ_n and ϕ_{n-1} are appropriate functions.

The following examples illustrate the separability condition.

Additive Returns

If returns are additive the value of state (m, i_m) under plan A_m is given by

$$f(m, i_m, A_m) = r(m, i_m, k_m) + \ldots + r(n, i_n, k_n) + \ldots + r(1, i_1, k_1) \tag{1.40}$$

$$= \sum_{n=1}^{m} r(n, i_n, k_n)$$

$$= r(m, i_m, k_m) + \sum_{n=1}^{m-1} r(n, i_n, k_n)$$

$$= r(m, i_m, k_m) + f(m-1, i_{m-1}, A_{m-1}). \tag{1.41}$$

Equation 1.41 is in the form of equation 1.38 so that the separability condition is satisfied.

Discounted Returns

If returns are discounted by a factor b the value of state (m, i_m) under plan A_m is given by

$$f(m, i_m, A_m) = r(m, i_m, k_m) + br(m-1, i_{m-1}, k_{m-1}) + \ldots$$
$$+ b^{m-n}r(n, i_n, k_n) + \ldots + b^{m-1}r(1, i_1, k_1) \qquad (1.42)$$
$$= \sum_{n=1}^{m} b^{m-n}r(n, i_n, k_n)$$
$$= r(m, i_m, k_m) + b \sum_{n=1}^{m-1} b^{m-1-n}r(n, i_n, k_n) \qquad (1.43)$$
$$= r(m, i_m, k_m) + bf(m-1, i_{m-1}, A_{m-1}) \qquad (1.44)$$

Equation 1.44 is in the form of equation 1.38 so the separability condition is satisfied.

Multiplicative Returns

Another value function which satisfies the separability condition is where the value of a state is given by the product of the returns.

$$f(m, i_m, A_m) = \prod_{n-1}^{m} r(n, i_n, k_n) \qquad (1.45)$$
$$= r(m, i_m, k_m) \prod_{n=1}^{m-1} r(n, i_n, k_n)$$
$$= r(m, i_m, k_m) f(m-1, i_{m-1}, A_{m-1}) \qquad (1.46)$$

Counter examples

Failure to satisfy the separability condition arises when for at least one plan the right hand side of equation 1.37 cannot be reduced to the form of equation 1.38. An example is the value function,

$$f(3, i_3, A_3) = r(3, i_3, k_3) \cdot r(2, i_2, k_2) + r(1, i_1, k_1) \cdot r(3, i_3, k_3)$$
$$+ r(2, i_2, k_2) \cdot (r1, i_1, k_1) \qquad (1.47)$$

The right hand side of equation 1.47 cannot be reduced to the form of the right hand side of equation 1.38; it cannot be rearranged into a function of $r(3, i_3, k_3)$ and $f(2, i_2, A_2)$ where $f(2, i_2, A_2)$ is a function of $r(2, i_2, k_2)$ and $r(1, i_1, k_1)$.

In many problems in the management area there is some degree of separability. Part of the art of dynamic programming is to make the maximum use of separability without going so far that the formulation becomes invalid. This is illustrated by the following elementary shortest path problem.

We wish to find the quickest means of travel from a cottage to a nearby village. The journey involves crossing a field to a gate and then going down a lane to the village. It takes 2 minutes to walk from the gate to the village and 1 minute to cycle from the gate to the village. It takes 4 minutes to cross the field pushing a bicycle and 1 minute to cross the field without a bicycle.

Suppose that we apply the value iteration method assuming additive returns in accordance with the network diagram of Figure 1.8.

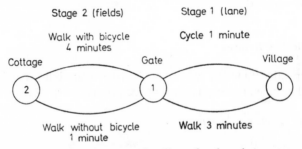

Figure 1.8. Getting to the village: first formulation

This gives the solution, walk without bicycle from 2 to 1, bicycle from 1 to 0. But it is impractical to cycle from 1 to 0 (owing to lack of bicycle) unless one has walked from 2 to 1 with the bicycle, and this solution is wrong. The error arises in the formulation. In terms of the separability condition the mistake can be expressed as follows.

Consider the one stage plan, 'cycle down the lane'. Let this be plan A_1^1. If the separability condition holds, the value of state 1 under plan A_1^1 will be expressible in the form,

$$f(1, A_1^1) = \Phi_1(r(1, k_1), f(0)) \tag{1.48}$$

But the value of plan A_1^1 depends on whether the bicycle was brought across the field. Thus $f(1, A_1)$ is a function of the action k_2 at stage 2. This variable does not appear in equation 1.48 and hence the problem is not separable in the current formulation.

A valid formulation requires the state description illustrated in Figure 1.9. There state (1,1) corresponds to being at the gate with the bicycle and state (1,2) to being at the gate without the bicycle. This reformulation eliminates the plan of walking across the field without the bicycle and cycling down the lane.

The formulation has been made valid by extending the state discription from the very simple form of Figure 1.8 to the less simple form of Figure 1.9. In more complex examples one finds similarly that the

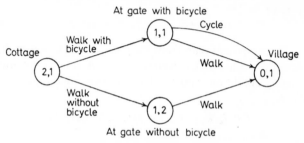

Figure 1.9. Getting to the village: second formulation

validity of dynamic programming as regards separability can be extended by increasing the number of states, but that this extension is achieved at the cost of increased computation.

1.10.2 Optimality Condition

Consider again the optimisation problem stated in equation 1.36. The separability condition requires that the value of each state can be calculated recursively *under any given plan*. In the value iteration method we not only calculate values recursively but also discard subplans at each state. For this to be valid the following condition must hold.

For every state and action, the optimal plans is to consist of the given action followed by the plan which is optimal with respect to the successor state.

This condition relates to the Principle of Optimality enunciated by Richard Bellman (Bellman 1957, 1962). We present it as a condition rather than a principle because it does not hold for every problem formulation. An algebraic statement of the optimality condition is now derived.

By the separability condition, for every n stage plan, A_n, we have

$$f(n, i_n, A_n) = \Phi_n(r(n, i_n, k_n), f(n-1, i_{n-1}, A_{n-1}))$$ (1.49)

Plan A_n consists of action k_n followed by the $n-1$ stage plan A_{n-1}. We can express this by the equation

$$A_n = k_n + A_{n-1}$$ (1.50)

We then have

$$f(n, i_n, A_n) = f(n, i_n, k_n + A_{n-1})$$ (1.51)

Let A_{n-1}^0 be an optimal plan from state $(n-1, i_{n-1})$, that is, from the successor state to state (n, i_n) given action k_n. Assuming maximisation, the optimality condition requires that for every state (n, i_n), action k_n and plan A_{n-1}

$$f(n, i_n, k_n + A_{n-1}^0) \geqslant f(n, i_n, k_n + A_{n-1})$$ (1.52)

Inequality 1.52 is the algebraic statement of the optimality condition The following examples illustrate the application of the condition.

Additive Returns

For additive returns under any plan,

$$f(n, i_n, k_n + A_{n-1}) = r(n, i_n, k_n) + f(n-1, i_{n-1}, A_{n-1})$$ (1.53)

Hence the optimality condition requires that

$$r(n, i_n, k_n) + f(n-1, A_{n-1}^0) \geqslant r(n, i_n, k_n) + f(n-1, A_{n-1})$$ (1.54)

$$f(n-1, A_{n-1}^0) \geqslant f(n-1, A_{n-1})$$ (1.55)

which is true by definition of plan A_{n-1}^0. Hence the optimality condition is satisfied.

Discounted and Multiplicative Returns

For discounted returns,

$$f(n, i_n, k_n + A_{n-1}) = r(n, i_n, k_n) + bf(n-1, i_{n-1}, A_{n-1}) \quad (1.56)$$

where b is a discount factor. The optimality condition requires that

$$r(n, i_n, k_n) + bf(n-1, i_{n-1}, A_{n-1}^0) \geqslant r(n, i_n, k_n) + bf(n-1, i_{n-1}, A_{n-1}).$$
$$(1.57)$$

Provided b is nonnegative, inequality 1.57 holds and the optimality condition is satisfied. By similar reasoning it can be shown that the value iteration method is valid for the optimisation of the product of nonnegative returns.

Counter example

Consider the value function

$$f(n, i_n, A_n) = r(n, i_n, k_n) + f(n-1, i_{n-1}, A_{n-1}) - (4 - f(n-1, i_{n-1}, A_{n-1}))^2$$
$$(1.58)$$

This could correspond to a situation where plan A_{n-1} yields a total profit $f(n-1, i_{n-1}, A_{n-1})$ over stages $n-1$ to 1, action k_n brings a return $r(n, i_n, k_n)$ which is added to the previous profit, but a tax has to be paid which is equal to the square of the difference between the previous total profit and some target quantity. The expression

$$(4 - f(n-1, i_{n-1}, A_{n-1}))^2$$

is the amount of tax. This interpretation assumes forward recurrence. We wish to maximise the value of state (n, i_n).

Consider further that plans $A_{n-1} = 1, 2, 3$ respectively yield the values $f(n-1, i_{n-1}, A_{n-1}) = 3, 4, 6$. The optimality condition requires that for all A_{n-1}

$$r(n, i_n, k_n) + f(n-1, i_{n-1}, A_{n-1}^0) - (4 - f(n-1, i_{n-1}, A_{n-1}^0))^2$$
$$\geqslant r(n, i_n, k_n) + f(n-1, i_{n-1}, A_{n-1}) - (4 - f(n-1, i_{n-1}, A_{n-1}))^2 \quad (1.59)$$

where A_{n-1}^0 is the optimal $n-1$ stage plan. Cancel the $r(n, i_n, k_n)$ terms

$$f(n-1, i_{n-1}, A_{n-1}^0) - (4 - f(n-1, i_{n-1}, A_{n-1}^0))^2$$
$$\geqslant f(n-1, i_{n-1}, A_{n-1}) - (4 - f(n-1, i_{n-1}, A_{n-1})) \quad (1.60)$$

Now the optimal $n-1$ stage plan is plan 3 which gives

$$f(n-1, i_{n-1}, A_{n-1}^0) = 6$$

The left hand side of inequality 1.60 therefore has the value

$$6 - (4 - 6)^2 = 2$$

However, for plan $A_{n-1} = 2$ we have $f(n-1, i_{n-1}, A_{n-1}) = 4$ which gives the right hand side of inequality 1.60 the value 4. Hence the inequality fails and the optimality condition is violated.

A practical interpretation of this counterexample is that with some forms of wealth tax it may not be optimal as regards the total return over n stages, to maximise the return over $n-1$ stages.

Theorem 1.1

The maximum

$$f(m, i_m) = \max_{A_m \in W_m} [\phi_m(r(m, i_m, k_m), \ldots, r(n, i_n, k_n), \ldots, r(1, i_1, k_1))] \quad (1.61)$$

can be found by the value iteration

$$f(n, i_n) = \max_{k_n \in K_n} [\Phi_n(r(n, i_n, k_n), f(n-1, i_{n-1}))] \quad (1.62)$$

carried out over all relevant states and actions from given terminal values, if and only if the separability and optimality conditions hold.

Proof

If the separability condition does not hold there will be one or more plans for which the equation

$$f(i_n, A_n) = \Phi_n(r(n, i_n, k_n), f(n-1, i_{n-1}, A_{n-1})) \quad (1.63)$$

is invalid. If such a relationship does not hold an algorithm based on it will not be generally valid. Hence the separability condition is necessary.

Now, by definition,

$$f(n, i_n) = \underset{A_n \in W_n}{\text{Max}} [f(n, i_n, A_n)] = f(n, i_n, A_n^0) \qquad (1.64)$$

Therefore

$$f(n, i_n, k_n) = \underset{A_{n-1} \in W_{n-1}}{\text{Max}} [f(n, i_n, k_n + A_{n-1})] \qquad (1.65)$$

Suppose the separability condition holds. If the optimality condition also holds then for all states (n, i_n), actions k_n and plans A_{n-1}

$$f(n, i_n, k_n + A_{n-1}^0) \geqslant f(n, i_n, k_n + A_{n-1}) \qquad (1.66)$$

therefore

$$f(n, i_n) = \underset{k_n \in K_n}{\text{Max}} [f(n, i_n, k_n + A_{n-1}^0)] \qquad (1.67)$$

$$= \underset{k_n \in K_n}{\text{Max}} [\Phi_n(r(n, i_n, k_n), f(n-1, i_{n-1}, A_{n-1}^0))] \qquad (1.68)$$

$$= \underset{k_n \in K_n}{\text{Max}} [\Phi_n(r(n, i_n, k_n), f(n-1, i_{n-1}))] \qquad (1.69)$$

But 1.69 is the value iteration equation. Hence the two conditions are together sufficient for the validity of the value iteration method.

If the optimality condition does not hold then for some plan A_{n-1}^1, say, inequality 1.66 fails. Application of plan A_{n-1}^1 for stages $n-1$ to 1 and action k_n at stage n will lead to a larger value for state (n, i_n) than will the plan A_{n-1}^0 found by the value iteration algorithm. Further iteration from a suboptimal n stage plan will not in general lead to the optimal m stage plan. Hence the optimality condition is necessary.

Hence the two conditions are necessary and sufficient for the validity of the value iteration method.

Q.E.D.

1.11 MINIMISATION OF THE MAXIMUM RETURN

The value iteration method can be used to find the path through a serial (or progressive) network, the longest stage of which is as short as possible. It can also be used to find the path whose shortest stage is as long as possible. These are known as the Minmax and Maxmin problems respectively. Examples of such problems are as follows; an aircraft is to be routed to a destination in such a way that it can take the maximum payload, the payload being restricted by the fuel required for the longest leg of the journey; a production process is to be arranged so that the maximum electric current taken at any time is as low as possible; the throughput of a production line is to be maximised, and is governed by the throughput of the slowest stage. A worked example of the latter application in a steel rolling mill is given by White 1969.

Consider the problem of finding the path through the network shown in Figure 1.10 whose longest stage is as short as possible. The problem

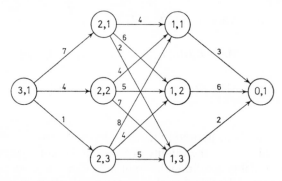

Figure 1.10. Serial network with returns; minmax and maxmin problems

is formulated in the same way as the shortest path problem, except that the value of a state under a given plan is now defined as the length of the longest stage which occurs in that plan. Thus the value $f(n, i_n, A_n)$ of state (n, i_n) under plan A_n, which comprises actions $k_n, k_{n-1}, \ldots, k_1$, is given by

$$f(n, i_n, A_n) = \text{Max} \, [r(n, i_n, k_n): r(n-1, i_{n-1}, k_{n-1}): \ldots : r(1, i_1, k_1)]$$

$$(1.72)$$

where the colon is to be read as 'or'. We see that

$$f(n, i_n, A_n) = \text{Max } [r(n, i_n, k_n): \text{Max } [r(n-1, i_{n-1}, k_{n-1}): \ldots$$
$$\ldots: r(1, i_1, k_1)]] \tag{1.73}$$

$$= \text{Max } [r(n, i_n, k_n): f(n-1, i_{n-1}, A_{n-1})] \tag{1.74}$$

Equation 1.74 is of the form of equation 1.38 and hence the separability condition is satisfied.

Equation 1.74 is equivalent to

$$f(n, i_n, k_n + A_{n-1}) = \text{Max } [r(n, i_n, k_n): f(n-1, i_{n-1}, A_{n-1})] \tag{1.75}$$

Since we are minimising, the optimality condition requires that, for all A_{n-1}

$$f(n, i_n, k_n + A_{n-1}^0) \leqslant f(n, i_n, k_n + A_{n-1}) \tag{1.76}$$

where A_{n-1}^0 is an optimal $n-1$ stage plan. Using equation 1.75 the condition becomes, for all A_{n-1}

$$\text{Max } [r(n, i_n, k_n): f(n-1, i_{n-1}, A_{n-1}^0)] \leqslant \text{Max } [r(n, i_n, k_n): $$
$$f(n-1, i_{n-1}, A_{n-1})] \tag{1.77}$$

Now since we are minimising,

$$f(n, i_n, A_n^0) \leqslant f(n, i_n, A_n) \tag{1.78}$$

As regards the right hand side of inequality 1.77 there are the two possibilities

$$r(n, i_n, k_n) \geqslant \underset{A_{n-1} \in W_{n-1}}{\text{Max}} [f(n-1, i_{n-1}, A_{n-1})] \tag{1.79}$$

$$r(n, i_n, k_n) < \underset{A_{n-1} \in W_{n-1}}{\text{Max}} [f(n-1, i_{n-1}, A_{n-1})] \tag{1.80}$$

For the former, inequality 1.77 holds because both sides take the value $r(n, i_n, k_n)$. For the second possibility inequality 1.77 becomes

$$\text{Max } [r(n, i_n, k_n): f(n-1, i_{n-1}, A_{n-1}^0)] \leqslant \underset{A_{n-1} \in W_{n-1}}{\text{Max}} [f(n-1, i_{n-1}, A_{n-1})] \tag{1.81}$$

which true in view of inequality 1.78. Hence the optimality condition holds, and the minmax path can be found by application of the value iteration equation,

$$f(n, i_n) = \underset{k_n \in K_n}{\text{Min}} [\text{Max} [r(n, i_n, k_n): f(n-1, i_{n-1})]]. \qquad (1.82)$$

The calculations and the optimal paths for the problems of finding Minmax and Maxmin paths through the network in Figure 1.10 are shown in Tables 1.15 to 1.18. The general recurrence relation for maximising the minimum return is

$$f(n, i_n) = \underset{k_n \in K_n}{\text{Max}} [\text{Min} [r(n, i_n, k_n): f(n-1, i_{n-1})]] \qquad (1.83)$$

Table 1.15. MINMAX PROBLEM: CALCULATIONS

Stage	State	Action	Trial Value
1	1	*1*	*3*
1	2	*1*	*6*
1	3	*1*	*2*
2	1	1	Max (4 : 3) = 4
2	1	2	Max(6 : 6) = 6
2	1	3	Max(2 : 2) = *2*
2	2	*1*	Max(4 : 3) = *4*
2	2	2	Max(5 : 6) = 6
2	2	3	Max(7 : 2) = 7
2	3	1	Max(8 : 3) = 8
2	3	2	Max(4 : 6) = 6
2	3	3	Max(5 : 2) = 5
3	1	1	Max(7 : 2) = 7
3	1	2	Max(4 : 4) = *4*
3	1	3	Max(1 : 5) = 5

Table 1.16. MINMAX PROBLEM: OPTIMAL PROCESS

Stage	State	Action	Value
3	1	2	4
2	2	1	4
1	1	1	3

Table 1.17. MAXMIN PROBLEM: CALCULATIONS

Stage	State	Action	Trial Value
1	1	1	*3*
1	1	1	*6*
1	3	1	*2*
2	1	1	Min(4 : 3) = 3
2	1	2	Min(6 : 6) = *6*
2	1	3	Min(2 : 2) = 2
2	2	1	Min(4 : 3) = 3
2	2	2	Min(5 : 6) = *5*
2	2	3	Min(7 : 2) = 2
2	3	1	Min(8 : 3) = 3
2	3	2	Min(4 : 6) = *4*
2	3	3	Min(5 : 2) = 2
3	1	*1*	Min(7 : 6) = *6*
3	1	2	Min(4 : 5) = 4
3	1	3	Min(1 : 4) = 1

Table 1.18. MAXMIN PROBLEM: OPTIMAL PROCESS

Stage	State	Action	Value
3	1	1	6
2	1	2	6
1	2	1	6

EXERCISES

Q1.1. The network of Figure 1.11 represents a group of cities through which a carrier can travel, using only the roads shown in the directions indicated. The figures shown against the roads represent the profit accruing from the corresponding journey. Find the path which maximises the carrier's profit.

Q.1.2. A company owns a warehouse in which it can store a certain commodity. At the beginning of each year the warehouse is either full

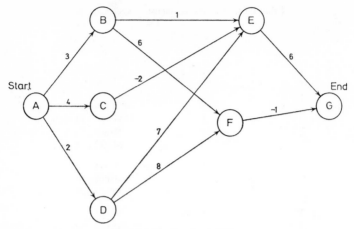

Figure 1.11. Carriers problem

or empty. If the warehouse is full the company can decide from among the following actions for the coming year:

1. Buy and sell nothing, leaving warehouse full at start of the next year.
2. Sell the contents of the warehouse and purchase fresh stock leaving the warehouse again full at the start of the next year.
3. Sell the contents and buy nothing.

If the warehouse is empty at the start of the year the company can either:

1. Buy nothing.
2. Buy sufficient to fill the warehouse.

A forecast of the net price of buying and selling the commodity for the next five years has been prepared and is as follows.

Year	1	2	3	4	5
Buying price	10	13	18	14	11
Selling price	11	13	15	15	11

If the warehouse is kept full for a year with no stock movement there is a charge of one unit. Determine the optimum buying and selling plan for the company over the five year period. Assume that the warehouse is full initially and that it is to be full at the end of the planning period.

Answer	Year:	1	2	3	4	5
	Action:	Sell and buy	Sell and buy	Sell	Do nothing	Buy

Ql. 3. Solve Ql. 2 assuming future costs are discounted with interest rate 25% p.a.

Answer	Year:	1	2	3	4	5
	Action:	Sell and buy	Sell	Do nothing	Do nothing	Buy

Q.1.4. Water supply for a new town can be provided by adding capacity to some sections of an existing network of pipes. The cost of adding the necessary capacity to each section is shown in Figure 1.12.

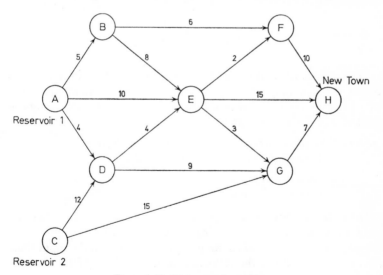

Figure 1.12. Water supply problem

4

Determine the cheapest way to arrange for the new supply. Arrows indicate directions of flow. The whole supply must come from either reservoir 1 or reservoir 2.

Answer Add capacity to links AD, DE, EG, GH.

Q.1.5. Metal initially in strips of thickness 10 cm is to be rolled into strips of thickness 4 cm by passing through three rolling stands in succession. The cost of operating a stand depends on the input thickness of the metal and the reduction in thickness required, as shown in Table 1.19. Blanks in the table indicate that the corresponding reduction is not feasible.

Table 1.19. COST OF ROLLING METAL

Input thickness (cm)	Reduction (cm)		
	1	2	3
10	4	6	8
9	4	6	9
8	4	7	12
7	4	10	—
6	5	10	—
5	7	12	—
4	10	—	—

Use a forward recurrence dynamic programming formulation to obtain the minimum cost sequence of reductions. Extend your solution to cover the case where a fourth stand is added and an output thickness of 3 cm is required.

Answer 1. Sequence of reductions is 3 cm, 1 cm, 2 cm.
Total cost 22 units.
2. Sequence of reductions is 3 cm, 1 cm, 1 cm, 2 cm.
Total cost 31 units.

Q.1.6. Metal initially in strips of thickness 5 cm is to be rolled into strips of thickness 1 cm by passing through a number of rolling stands in succession. The cost of running a stand depends on the input thickness of the metal and the reduction in thickness required, the costs being as shown in the Table 1.20.

Table 1.20. COST OF ROLLING METAL

Input thickness (cm)	Reduction (cm)		
	1	2	3
5	7	9	15
4	6	8	—
3	6	7	—
2	4	—	—

Determine the number of stands and sequence of reductions which minimises the cost of rolling.

Answer: Sequence of reductions is 2 cm, 2 cm.
Two stands are used.

Q.1.7. The payload of an aircraft on a certain journey is limited by the quantity of fuel which must be carried on the longest stage of the journey. Given the network of distances between airports shown in Figure 1.13 find the route which maximises the pay load which can be carried.

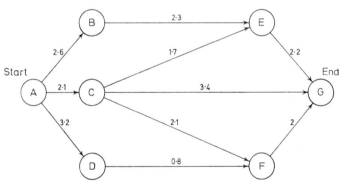

Figure 1.13. Distances between airports (thousands of miles)

Q1.8. A company plans to construct three new buildings A, B, C, at a rate of one per year, and it can choose the order in which they are built. The cost of constructing a building depends on which other buildings have already been completed in accordance with Table 1.21.

4*

Table 1.21.

Already built	Construction cost		
	A	B	C
Nothing	10	8	6
A	—	9	8
B	13	—	9
C	11	10	—
A and B	—	—	11
A and C	—	12	—
B and C	14	—	—

For tax purposes it is best to arrange the sequence of building so that the smallest of the annual costs is as large as possible. In what sequence should the buildings be constructed? Determine also the sequence which minimises total cost.

Answer: Maxmin optimal sequence is A, B, C.
Minimum cost sequence is C, A, B.

CHAPTER 2

Deterministic Applications

2.1 INTRODUCTION

In deterministic decision processes, the result of every action is known with certainty. In dynamic programming terms this means that the returns and the transitions are uniquely determined once a particular plan is chosen. The problems discussed in Chapter 1 were deterministic. In Chapter 3 we introduce stochastic problems, where a particular plan determines the probabilities that various returns will be generated and various transitions will occur. In this chapter we consider applications of the value iteration method to deterministic problems.

The applications of the value iteration method range over a very broad area. Problems which involve a sequence of decisions in time are obvious candidates. These arise in such fields as production planning, stock control, replacement and investment decision making. Here a stage in a dynamic programming sense is identifiable with a time interval such as a week, month or year. Other applications may involve a sequence of decisions not directly related to time. This includes sequential production processes, optimal path problems and search problems. A third area is that of allocation problems, which involve the allocation of one or more limited resources to various uses. A stage involves the allocation of some quantity of a particular resource to a particular use, and a problem can be thought of as a sequence of such allocations. A fourth area is combinatorial and graph theoretic problems, including scheduling, sequencing and set partitioning.

43

Particular applications may not fit neatly into any of the areas just mentioned, or may involve more than one area. For example, a problem may involve a set of time sequential decisions followed by a set of space sequential decisions. The key feature of a value iteration application is that it must be identifiable as a serial or progressive directed network problem. However, value iteration may not be the only technique relevant to a given problem, but may give way to, or be used in conjunction with such techniques as the calculus, integer programming or branch and bound.

Before going on to consider particular applications we summarise the general approach to the formulation of problems. We follow this by the detailed solution of a production planning problem a replacement problem, a set partitioning problem and two resource allocation problems. In the course of solving these we consider some points of methodology, namely the treatment of constraints and of problems with multidimensional state space.

2.2 FORMULATION PROCEDURE

In formulating problems for solution by dynamic programming it is advisable to follow a standard procedure. Some elements of this have already been introduced in Chapter 1, in particular the defining of the terms stage, state, action etc. However, the shortest path problem was deliberately chosen for its simplicity and many features of it could be regarded as obvious whereas in other problems these same features require definition. The following procedure for problem formulation is suggested.

1. List the terms *stage, state, action, return, optimal value of a state* and against each write down the interpretation of that term in the current problem, together with the symbol used to denote the relevant quantity. In case of difficulty, try to draw a network diagram.
2. Write down the following equations; some of these have been encountered already and the remainder will be exemplified shortly,
   a) Recurrence relation
   b) Transition equation
   c) Return equation

d) Inequalities determining the range of the stage, state and action variables
e) Terminal values.
3. Check that the validity conditions are satisfied. If the returns are additive, discounted, multiplicative or of the minmax or maxmin variety it is sufficient to check that the returns are independent.

It is not unusual to find that a problem can be formulated in more than one way and part of the art of dynamic programming lies in deciding the most efficient formulation for the problem on hand. Once the formulation along the above lines has been completed it is as well to check for any special rules or short cuts which might make the formulation more efficient. One of the advantages of following a formal procedure is that every term and condition is defined and the scope and effect of possible modifications are more readily apparent.

2.3 PRODUCTION PLANNING PROBLEM

The following example illustrates the formulation procedure and also the use of value iteration for dynamic production planning.

A boat builder has the orders shown in Table 2.1 for delivery at the end of the months indicated. He can build up to four boats in any month and can additionally hold up to three boats in stock. If boats are built in a particular month there is an overhead cost of £4000 plus a construction cost of £10 000 per boat. Stock holding costs

Table 2.1. BOATBUILDERS ORDERS

Month	Feb	Mar	Apr	May	Jun	Jul
No. of boats	1	2	5	3	2	1

are £1000 per boat per complete month. In what months should boats be built and in what quantities if costs are to be minimised? Assume that the stock is zero initially and is to be zero after the July delivery, and that orders must be met.

2.3.1 Formulation

The first step is to identify how the problem fits into the framework of stages, states, actions, etc. It is a useful aid at this point to imagine the form the final process will take. Thus, in each month we shall start with some stock level and then decide to build some number of boats. Stockholding, overhead and construction costs may be incurred. At the end of the month a number of boats will be delivered, and we shall then enter the next month with a new stock level and the cycle of activities will be repeated. The dynamic programming formulation is as follows.

Each month corresponds to a *stage*. Let n be the number of months or stages which remain until the planning horizon, that is the end of July, is reached. Let a *state* (n, i) be defined by the number of boats i which are in stock at the beginning of the month corresponding to stage n. Thus state $(4, 2)$ corresponds to 2 boats in stock at the beginning of April. Let *action* k at a given state be to build k boats in the current month. The *return* in any month is the cost incurred in that month. The return will depend on the stock level at the start of the month and the number of boats built, and is denoted by $r(i, k)$. The costs of building and storage are the same in every month, so the stage variable does not appear in the return function. However, if costs did vary with the month we could allow for this and would then denote the return function by $r(n, i, k)$. The *optimal value of a state* is the sum of the costs incurred when the system starts in that state and an optimal plan is followed. The optimal value of state (n, i) is denoted by $f(n, i)$. This formulation is summarised in Table 2.2.

The recurrence relation follows immediately once the leading terms have been defined. It is

$$f(n, i) = \underset{k \in K}{\text{Min}} \; [r(i, k) + f(n-1, j)] \qquad (2.1)$$

where k is a member of the set k of feasible actions in state (n, i), and $(n-1, j)$ is the successor state given a current state (n, i) and action k. The state variable j in the successor state $(n-1, j)$ is determined by the transition equation,

$$j = i + k - d \qquad (2.2)$$

where d is the demand at the current stage. Equation 2.2 corresponds to the fact that the stock at the succeeding stage is given by the current stock plus the build quantity minus the demand.

The return equation determines the return $r(i, k)$ associated with stock level i and build quantity k. It is convenient to work in units of £1000. The total construction cost of £140 000 must be met however the production is scheduled and need not be considered in determining the optimal schedule. If i boats are in stock at the start of the month and none are built there is only a storage cost of i units. If k boats are built there is additionally an overhead cost of 4 units. Hence the returns are given by

$$\left. \begin{aligned} r(i, 0) &= i \\ r(i, k) &= 4+i \quad k > 0 \end{aligned} \right\} \tag{2.3}$$

Table 2.2. BOATBUILDER'S PROBLEM: SPECIFICATION

Stage	A month	n
State	Stock level i at start of month when n months remain	(n, i)
Action	Build k boats in current month	k
Return	Cost in current month	$r(i, k)$
Optimal value of a state	Total cost when the system starts in state (n, i) and an optimal plan is followed	$f(n, i)$

The next step is to define the sets of possible values or 'space' of the state, stage and action variables. In the optimal path problems of Chapter I these sets were evident from the network diagrams. In the current example the stage variable ranges from $n = 0$ to 6. The state variable, which we may indicate by i when the stage number is immaterial or by i_n when we wish to emphasise that we are considering a value at stage n, is constrained as follows. Firstly the stock level must not exceed 3

$$0 \leqslant i \leqslant 3 \tag{2.4}$$

Secondly the stock must not exceed the sum of the remaining demands

$$i_n \leqslant \sum_{w=1}^{n} d_w \tag{2.5}$$

where d_w is the demand at stage w. Thirdly the stock level at the start of any month must be no more than four boats below the demand in that month

$$i_n \geqslant d_n - 4 \qquad (2.6)$$

The action variable denoted k, or k_n when stage n is particularly intended, is constrained as follows. Firstly, up to four boats can be built in a month

$$0 \leqslant k \leqslant 4 \qquad (2.7)$$

Secondly the build quantity must not be such as to make the subsequent stock level exceed either the maximum of 3 or the remaining demand

$$\left. \begin{aligned} k_n &\leqslant 3 + d_n - i_n \\ k_n &\leqslant \sum_{w=1}^{n} d_w - i_n \end{aligned} \right\} \qquad (2.8)$$

Thirdly, demands must be met

$$k_n \geqslant d_n - i_n \qquad (2.9)$$

Finally, the initial and terminal states are known precisely

$$i_0 = i_6 = 0 \qquad (2.10)$$

If there is only one terminal state, as here, we can choose its value arbitrarily and it is convenient to put it to zero. However, the technique extends readily to the case where there are several possible terminal states, provided the relative values of these states are known. Thus if the value associated with having 0, 1, 2, etc., boats in stock at the end of July were known we could optimise over these possible terminal states. For the current problem we put $f(0, 0) = 0$.

The returns are additive and independent and so the validity conditions are satisfied. In a less heavily constrained production planning problem a progressive formulation may reduce the computation, see exercise Q2.1.

2.3.2 Calculation

The calculation is shown in detail in Table 2.3. At stage 1 the remaining demand is for one boat so that from inequality 2.8 we need only consider states (1, 0) and (1, 1). In other words the stock level at stage

Table 2.3. BOATBUILDER'S PROBLEM: CALCULATIONS

Stage	State	Action	Trial Value	Demand
1	0	*1*	*4*	1
1	1	*0*	*1*	
2	0	2	$4+4 = 8$	2
2	0	*3*	$4+1 = 5$	
2	1	1	$5+4 = 9$	
2	1	2	$5+1 = 6$	
2	2	*0*	$2+4 = 6$	
2	2	1	$6+1 = 7$	
2	3	*0*	$3+1 = 4$	
3	0	*3*	$4+5 = 9$	3
3	0	4	$4+6 = 10$	
3	1	2	$5+5 = 10$	
3	1	3	$5+6 = 11$	
3	1	4	$5+6 = 11$	
3	2	1	$6+5 = 11$	
3	2	2	$6+6 = 12$	
3	2	3	$6+6 = 12$	
3	2	*4*	$6+4 = 10$	
3	3	*0*	$3+5 = 8$	
3	3	1	$7+6 = 13$	
3	3	2	$7+6 = 13$	
3	3	3	$7+4 = 11$	
4	1	*4*	$5+9 = 14$	5
4	2	*3*	$6+ 9 = 15$	
4	2	4	$6+10 = 16$	
4	3	*2*	$7+9 = 16$	
4	3	3	$7+10 = 17$	
4	3	4	$7+10 = 17$	
5	0	*3*	$4+14 = 18$	2
5	0	4	$4+15 = 19$	

Stage	State	Action	Trial Value	Demand
5	1	2	$5+14 = 19$	
5	1	3	$5+15 = 20$	
5	1	4	$5+16 = 21$	
5	2	1	$6+14 = 20$	
5	2	2	$6+15 = 21$	
5	2	3	$6+16 = 22$	
5	3	0	$3+14 = 17$	
5	3	1	$7+15 = 22$	
5	3	2	$7+16 = 23$	
6	0	1	$4+18 = 22$	1
6	0	2	$4+19 = 23$	
6	0	3	$4+20 = 24$	
6	0	4	$4+17 = 21$ ✓	

1 will not exceed one boat. Also the action at each of these states is completely determined by the constraints. At state (1, 0) one boat must built to meet demand, but no more or some will be left over. At state (1, 1) no boats must be built so that none shall be left over. Hence the values of these states are given by

$$\left. \begin{array}{l} f(1, 0) = r(0, 1) = 4 \\ f(1, 1) = r(1, 0) = 1 \end{array} \right\} \tag{2.11}$$

At stage 2 the state space is governed only by the maximum stock constraint, inequality 2.4. In state (2, 0), in view of the requirements not to overproduce (inequality 2.8) and to meet demand (inequality 2.9), the only possible actions are to build 2 or 3 boats. For $k = 2$ the test quantity is

$$r(0, 2)+f(1, 0) = 4+4 = 8 \tag{2.12}$$

and for $k = 3$,

$$r(0, 3)+f(1, 1) = 4+1 = 5 \tag{2.13}$$

Action $k = 3$ gives the smaller value and is therefore optimal, and $f(2, 0) = 5$.

At state (2, 1) the constraints leave actions $k = 1$, $k = 2$ available. The trial values under these actions are

$$r(1, 1)+f(1, 0) = 5+4 = 9 \atop r(1, 2)+f(1, 1) = 5+1 = 6 \Bigg\} \tag{2.14}$$

Action $k = 2$ is optimal and $f(2, 1) = 6$.

At state (2, 2) actions $k = 0$, $k = 1$ are available and evaluation similar to that for states (2, 0) and (2, 1) reveals action $k = 0$ to be optimal, and gives $f(2, 2) = 6$. At state (2, 3) only action $k = 0$ is feasible and $f(2, 3) = 4$.

Similar optimisation over stages 3 to 6 yields the results shown in Table 2.3, where as usual the optimal values and actions are italicised. At stage 4, state (4, 0) need not be considered because of inequality 2.6. At stage 6 it is only necessary to consider state (6, 0) because the initial stock level is zero.

The optimal process is picked out from Table 2.3 by entering at the initial state (6, 0) and noting the optimal action, namely $k = 1$. Application of the transition equation for this state and action determines the successor state,

$$j = i+k-d \atop = 0+4-1 = 3 \Bigg\} \tag{2.15}$$

Hence the next state is (5, 3). Here the optimal action is $k = 0$. Further application of the transition equation shows the next state to be (4, 1). Continuation through the table yields the solution shown in Table 2.4.

Table 2.4. BOATBUILDERS PROBLEM: OPTIMAL PROCESS

Stage	State	Action	Value
6 (Feb)	0	4	21
5 (Mar)	3	0	17
4 (Apr)	1	4	14
3 (May)	0	3	9
2 (Jun)	0	3	5
1 (Jul)	1	0	1

2.4 CONTROL

Suppose that the production or delivery quantity in the current month differs from that expected but that future orders and costs remain unchanged. To take a specific instance, suppose that two of the boats ordered for April are cancelled at the last minute. The stock level at the beginning of May will now be 2 boats and not zero as anticipated, so that the system is in state (3, 2). Table 2.3 shows that in this state the optimal action is to build 4 boats. We shall then have three boats in stock at the start of June, and Table 2.3 shows that we should then build zero boats in June, and again zero boats in July. Having once departed from the original optimal process, that process becomes irrelevant and to attempt to return to the original stock or production quantities merely because they were formerly optimal would be wrong.

Changes from month to month often involve not only the immediate stock level but also the pattern of future orders and possibly the costs of stock holding and production. It then becomes necessary to repeat the calculations with new data. If we know that the same calculation is going to recur frequently with different data it will be worthwhile to write a computer programme for it, or to use a standard package programme, see Appendix. It is then easy to repeat the calculation at intervals, feeding in the current demands, costs, constraints, stock level and terminal values. The planning horizon can be moved forward so that the planning period is maintained at an appropriate length.

2.5 CONSTRAINTS

In the boatbuilders problem the state and action space are restricted by equations and inequalities 2.4–2.10. Constraints which restrict only the state or action space, or both, are advantageous in dynamic programming because they reduce the amount of computation. By contrast, state and action space constraints can cause considerable procedural difficulties for other optimisation techniques. On the other hand, some constraints have the effect of requiring extension of the state space, as happened with the restriction on the use of the bicycle in the 'getting to the village' problem in Chapter I.

2.6 STATIONARITY

In the boatbuilder's problem the demand for boats varies from month to month. The problem is nonstationary with respect to demand. The cost function is the same at each stage and the problem is therefore stationary with respect to returns. A problem which is stationary with respect to states, actions and returns is a stationary problem.

One of the advantages of dynamic programming is its ability to solve nonstationary as well as stationary problems. Readers familiar with the economic batch quantity formula which is used in production planning and stock control will realise that the derivation of this formula assumes the problem to be stationary. If a problem is not stationary the repeated use of the economic batch quantity formula for differing demand levels will not in general give the optimal solution.

The optimal plan for a stationary problem will not always be stationary. However, interest often centres on finding the best stationary plan, particularly when the planning horizon is remote. The determination of optimal stationary plans in infinite stage problems is discussed in Chapters 4 and 5

2.7 REPLACEMENT PROBLEM

A machine in a production plant is inspected annually and either overhauled or replaced. The cost of overhaul and the scrap value of the machine are related to its age as shown in Table 2.5.

Table 2.5. REPLACEMENT PROBLEM: DATA

Age (years)	1	2	3	4
Cost of overhaul (£1 000 s)	7	3	9	—
Scrap value (£1 000 s)	10	5	2	0

The cost of a new machine is £20 000. The remaining life of the whole plant is 5 years, at the end of which period the then current machine is scrapped. The present machine is 3 years old next birthday. Determine a minimum cost replacement plan.

2.7.1 Formulation

Each year is a stage and n denotes the number of years remaining to the planning horizon. State (n, i) corresponds to having a machine of age i when n stages remain. Let action $k = 1$ be to replace a machine and action $k = 2$ be to overhaul it. The return associated with state (n, i) and action k is denoted by $r(n, k)$. The return function is stationary, and in fact the problem is a stationary problem. The optimal value of state (n, i), denoted $f(n, i)$, is the total cost when the system

Table 2.6. REPLACEMENT PROBLEM: SPECIFICATION

Stage	A year	n
State	Machine age i when n years remain	(n, i)
Action	1 = Replace 2 = Overhaul	k
Return	Cost associated with current state and action	$r(i, k)$
Optimal value of a state	Total cost when the system starts in that state and an optimal plan is followed	$f(n, i)$

starts in state (n, i) and an optimal plan is followed. This specification is summarised in Table 2.6.

The recurrence relation is

$$f(n, i) = \underset{k \in K_i}{\text{Min}} \, [r(i, k) + f(n-1, j)] \tag{2.16}$$

K_i is the set of possible actions in state i and is

$$\left. \begin{array}{ll} K_i = \{1, 2\}, & i = 1, 2, 3 \\ K_i = \{1\}, & i = 4 \end{array} \right\} \tag{2.17}$$

The transition rules are

$$\left. \begin{array}{ll} j = 1 & \text{if} \quad k = 1 \\ j = i+1 & \text{if} \quad k = 2 \end{array} \right\} \tag{2.18}$$

The returns are considered in £1 000 units and are as follows. If a one year old machine is replaced there is an acquisition cost of 20 units

Table 2.7. REPLACEMENT PROBLEM: CALCULATIONS

Stage	State	Action	Trial Value
0	1		-10
0	2		-5
0	3		-2
0	4		0
1	1	*1*	$10-10 = 0$
1	1	2	$7-5 = 2$
1	2	1	$15-10 = 5$
1	2	2	$3-2 = 1$
1	3	*1*	$18-10 = 8$
1	3	2	$9+ 0 = 9$
1	4	*1*	$20-10 = 10$
2	1	1	$10+ 0 = 10$
2	1	2	$7+ 1 = 8$
2	2	1	$15+0 = 15$
2	2	2	$3+8 = 11$
2	3	*1*	$18+0 = 18$
2	3	2	$9+10 = 19$
2	4	*1*	$20+0 = 20$
3	1	*1*	$10+8 = 18$
3	1	2	$7+11 = 18$ (tie)
3	2	1	$15+8 = 23$
3	2	2	$3+18 = 22$
3	3	*1*	$18+8 = 26$
3	3	2	$9+20 = 29$
3	4	*1*	$20+8 = 28$
4	1	*1*	$10+18 = 28$
4	1	2	$7+22 = 29$
4	2	1	$15+18 = 33$
4	2	2	$3+26 = 29$
4	3	*1*	$18+18 = 36$
4	3	2	$9+28 = 37$
4	4	*1*	$20+18 = 38$
5	3	*1*	$18+28 = 46$
5	3	2	$9+38 = 47$

for a new machine, from which the 10 units realized as scrap value of the old machine must be deducted. Thus

$$r(1,1) = 20 - 10 = 10$$

If a one year old machine is overhauled the return (cost) is 7 units, $r(1, 2) = 7$. By similar reasoning we get the remaining values of $r(i, k)$, which are given below in the form of a return matrix, $\mathbf{R}$.

$$\mathbf{R} = [r(i, k)] = \begin{bmatrix} 10 & 7 \\ 15 & 3 \\ 18 & 9 \\ 20 & M \end{bmatrix} \tag{2.19}$$

The symbol M indicates a large number.

The terminal values are determined by the scrap values finally realized. If the machine which is left when the planning horizon is reached is one year old it has a scrap value of 10 units. Since we are minimising positive costs, this corresponds to a terminal value of -10 units. The symbol $\mathbf{f}(n)$ indicates a column vector of state values at stage n, and the symbol $\mathbf{f}'(n)$ denotes the same values transposed to a row vector. The terminal values are,

$$\mathbf{f}'(0) = [f(0, i)] = [-10 \quad -5 \quad -2 \quad 0] \tag{2.20}$$

2.7.2 Calculation

The full table of calculations is shown in Table 2.7, and the optimal process is shown in Table 2.8. States (4, 2) and (4, 3) are inaccessible, but are included in Table 2.7. When calculations are carried out by computer it may be less trouble to include some irrelevant states than to arrange for their exclusion.

Table 2.8. REPLACEMENT PROBLEM: OPTIMAL PROCESS

Stage	State	Action	Value
5	3	1 (Replace)	46
4	1	1 (Replace)	28
3	1	1 (Replace)	18
2	1	2 (Overhaul)	8
1	2	2 (Overhaul)	1

2.8 ORDERED SET PARTITIONING PROBLEM

In an ordered set partitioning problem there is a set U of sequentially numbered elements,

$$U = \{1, 2, \ldots, m\}.$$

The set U is to be divided into subsets of sequentially numbered elements which cover U. A general subset, $X(i, j)$, $i \geqslant j$, consists of elements $i, i-1, \ldots, j$. The cost of forming subset $X(i, j)$ is known for all i and j. This type of problem is a one of a more general family of set partitioning problems in which the elements are not sequentially numbered. The following is an example.

A company receives orders for metal sheet in various thicknesses, denoted $1, 2, \ldots, m$. The company can meet the orders by making some thicknesses as 'standard' and then producing nonstandard thicknesses by cold rolling. The cost of making thicknesses i to j inclusive by making i as standard and the remainder by cold rolling of i is $r(i, j)$, $i \geqslant j$. Formulate a dynamic programme to determine which thicknesses should be made as standard if costs are to be minimised. Use the formulation to solve the problem with the data given in Table 2.9.

Table 2.9. ORDERED SET PARTIONING PROBLEM:
COST DATA, $r(i, j)$

j = smallest size in subset

		1	2	3	4
	1	5			
i = standard	2	7	5		
size	3	8	7	6	
	4	15	10	8	6

2.8.1 Formulation

One feasible solution would be to make all four sizes as standard. From Table 2.9 we see that this would cost $5+5+6+6 = 22$ units. Another solution is to make size 4 as standard and all the others by cold rolling and this would cost 15 units. In between these extremes there may be cheaper solutions. In dynamic programming terms the problem is progressive with only one state in each stagewise subset. The network diagram is shown in Figure 2.1. State (i) corresponds to the situation where sheet thicknesses i and below remain to be considered. Only a single variable is needed to define a state. Action

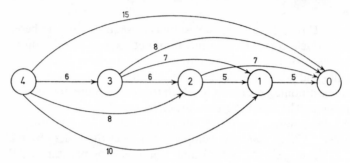

Figure 2.1. Ordered set partitioning problem: network diagram

k consists in making size k as the next smaller standard thickness, sizes $i-1$ to $k+1$ then being made by cold rolling of i. The formulation is summarised in Table 2.10.

Table 2.10. ORDERED SET PARTITIONING PROBLEM: SPECIFICATION

State	Situation where thicknesses i and below remain to be considered	(i)
Stage	Passing from one standard thickness to the next below	
Action	Make k as next smaller standard thickness	k
Return	Cost of making thicknesses i to $k+1$ inclusive from i as standard	$r(i, k+1)$
Optimal value of a state	Total cost when i is standard and an optimal plan is used for smaller sizes	$f(i)$

The recurrence relation is

$$f(i) = \underset{k<i}{\text{Min}} \left[r(i, k+1) + f(k) \right] \tag{2.21}$$

Table 2.11. ORDERED SET PARTITIONING
PROBLEM: CALCULATIONS

State	Action	Trial Value
0		*0*
1	0	*5*
2	*0*	*7*
2	1	5 + 5 = 10
3	*0*	*8*
3	1	7 + 5 = 12
3	2	6 + 7 = 13
4	0	15
4	1	10 + 5 = 15
4	2	8 + 7 = 15
4	*3*	6 + 8 = *14*

The terminal value is $f(0) = 0$. The calculations are shown in Table 2.11 and the optimal solution in Table 2.12. The optimal standard sizes are 4 and 3 and the total cost is 14 units.

Table 2.12. ORDERED SET PARTITIONING
PROBLEM: OPTIMAL PROCESS

State	Action	Value
4	3	14
3	0	8

2.9 ALLOCATION PROBLEMS

In an allocation problem a limited supply of one or more resources, such as raw materials, capital, labour, time, space, carrying capacity, etc., is to be allocated between a number of uses in such a way as to optimise some objective function. Two examples are presented and the general allocation problem is then discussed.

2.9.1 Allocation of one Resource

A factory can produce goods of types A, B and C in various quantities. Each product needs a raw material of which only four tons are available. Allocation of a certain quantity of the raw material to a certain product results in a corresponding return, as detailed in Table 2.13. The possible allocations are restricted to the levels shown in the table. Determine the allocation which maximizes the total return.

The dynamic programming formulation of the problem is as follows. A *stage* corresponds to the allocation of raw material to the production of a particular type of product. There are three products and hence three stages. When n types of product remain to be considered we are at stage n. The order in which products are considered can be chosen

Table 2.13. RETURNS RESULTING FROM VARIOUS
ALLOCATIONS

Allocation of raw material (tons)	Type of product		
	A	B	C
0	0	0	0
1	10	6	8
2	17	17	11
3	19	—	—

arbitrarily, and once it is chosen the product to be considered at stage n is referred to as product n. Products A, B, C will be considered at stages 3, 2, 1 respectively.

A state (n, i) corresponds to a situation where i tons of raw material remain to be allocated and n products remain to be considered. Action k corresponds to the allocation of k tons of raw material to the current product. The return resulting from the allocation of k tons of raw material to the current product is $r(n, k)$. The optimal value of a state is the total return generated when the system starts in that state and an optimal plan is followed, and is denoted by $f(n, i)$. This specification is summarised in Table 2.14.

Table 2.14. ALLOCATION OF ONE RESOURCE: PROBLEM SPECIFICATION

Stage	Allocation of raw material to a product. n = no. of products remaining	n
State	i tons of raw material remain to be allocated and n products remain to be considered	(n, i)
Action	Allocate k tons of raw material to current product	k
Return	Return from allocation of k tons of raw material to product n	$r(n, k)$
Optimal value of a state	Total return when state (n, i) is the starting state and an optimal plan is followed	$f(n, i)$

The recurrence relation is

$$f(n, i) = \operatorname*{Max}_{k \in K} [r(n, k) + f(n-1, j)] \qquad (2.22)$$

where k is the set of feasible allocations associated with state (n, i)

The transition equation is

$$j = i - k \qquad (2.23)$$

which expresses the fact that allocation of k tons at the current stage leaves $i-k$ tons available at the next stage. The recurrence relation 2.22 and transition equation can be combined into the recurrence relation 2.24

$$f(n, i) = \operatorname*{Min}_{k \in K} [r(n, k) + f(n-1, i-k)] \qquad (2.24)$$

The return function $r(n, k)$ is given by Table 2.13. The stage space is $0 \le n \le 3$. The state space is given by $0 \le i \le 4$ and the action space by

$$\left. \begin{array}{l} 0 \le k_1 \le 2 \\ 0 \le k_2 \le 2 \\ 0 \le k_3 \le 3 \\ k \le i \end{array} \right\} \qquad (2.25)$$

The first three inequalities in 2.25 reflect limitations on allocation implied by Table 2.13. The last constraint follows from the fact that no more than the remaining quantity of raw material can be allocated.

If raw material left over has no value (which we assume to be the case) the terminal values are $f(0, i) = 0$ for all i. The returns are additive and depend only on the current stage and action so that the validity conditions are satisfied.

2.9.2 Calculation

The calculations are shown in Table 2.15. At stage 1 the trial values under action k are given simply by $r(1, k)$ and follow directly from Table 2.13. The largest possible allocation is optimal at each state and the optimal values are

$$f(1, 0) = 0, f(1, 1) = 8, f(1, 2) = f(1, 3) = f(1, 4) = 11 \quad (2.26)$$

To illustrate the calculation at stage 2 consider state $(2, 1)$. One ton of raw material is available. We can either allocate nothing to product B or allocate the one ton. In the former case there is no income at stage 2 and the system moves to state $(1, 1)$, that is the one ton of raw material remains available at stage 1. The optimal value of state $(1, 1)$ has already been determined (equation 2.26 and Table 2.15) and is $f(1, 1) = 8$. Thus for action $k = 0$ the trial value of state $(2, 1)$ is

$$r(2, 0)+f(1, 1) = 0+8 = 8 \quad (2.27)$$

If the one ton is allocated to product B there will be an immediate return $r(2, 1) = 6$ units, and the system will move to state $(1, 0)$, corresponding to zero raw material at stage 1. The trial value of state $(2, 1)$ under action $k = 1$ is therefore

$$r(2, 1)+f(1, 0) = 6+0 = 6 \quad (2.28)$$

The set of feasible actions is now exhausted. Comparison of the trial values shows that action $k = 0$ is optimal and hence $f(2, 1) = 8$. This calculation and result are shown in Table 2.15, rows 15 and 16. Similar optimisation over the remaining states at stages 2 and 3 completes Table 2.15. At stage 3 it is only necessary to consider state $(3, 4)$ since the initial quantity available is known to be 4 tons. The optimal allocation is determined by following through the optimal process in Table 2.15. The optimal process is shown in Table 2.16. The solution is to allocate 1 ton to product A, 2 tons to product B and 1 ton to product C. The total return is then 35 units.

Table 2.15. ALLOCATION OF ONE RESOURCE: CALCULATIONS

Stage	State	Action	Trial Value
0	i		0
1	0	0	0
1	1	0	0
1	1	1	8
1	2	0	0
1	2	1	8
1	2	2	11
1	3	0	0
1	3	1	8
1	3	2	11
1	4	0	0
1	4	1	8
1	4	2	11
2	0	0	0
2	1	0	$0+8 = 8$
2	1	1	$6+0 = 6$
2	2	0	$0+11 = 11$
2	2	1	$6+ 8 = 14$
2	2	2	$17+ 0 = 17$
2	3	0	$0+11 = 11$
2	3	1	$6+11 = 17$
2	3	2	$17+ 8 = 25$
2	4	0	$0+11 = 11$
2	4	1	$6+11 = 17$
2	4	2	$17+11 = 28$
3	4	0	$0+28 = 28$
3	4	1	$10+25 = 35$
3	4	2	$17+17 = 34$
3	4	3	$19+ 8 = 27$

Table 2.16. ALLOCATION OF ONE RESOURCE: OPTIMAL PROCESS

Stage	State	Action	Value
3	4	1	35
2	3	2	25
1	1	1	8

2.9.3 Allocation of Several Resources

A factory can produce goods of types A, B and C in varying quantities. Each product needs two raw materials X and Y and only four tons of each material are available. The amounts of X and Y used up in producing 1, 2 or 3 items of products A, B, C are shown in Table 2.17.

Table 2.17. TWO RESOURCE ALLOCATION PROBLEM: REQUIREMENTS FOR RAW MATERIALS X AND Y

| | | | Type of product | | | |
| Quantity produced | A | | B | | C | |
	X	Y	X	Y	X	Y
1	1	2	1	2	1	1
2	2	2	2	3	1	2
3	3	3	2	3	2	3

The income resulting from the production of 1, 2, 3 items of products A, B, C is shown in Table 2.18. How much of each product should be produced to maximise total income?

Table 2.18. TWO RESOURCE ALLOCATION PROBLEM: INCOME LEVELS

| | Type of product | | |
Quantity produced	A	B	C
1	10	6	5
2	17	12	8
3	19	17	11

2.9.4 Formulation

The dynamic programming formulation is broadly similar to that used in the one resource allocation problem. A stage corresponds to the allocation of raw material to a product. When n types of product remain to be considered we are at stage n. The order in which products are considered is again immaterial and products A, B, C will again be considered at stages 3, 2, 1 respectively.

A state (n, x, y) corresponds to there being n products still to be considered with x tons of raw material X and y tons of raw material Y remaining. There are now two state variables x and y, and the problem is said to have two dimensional state space. From a computational point of view a problem with two or more dimensions of state space does not differ fundamentally from a problem with one dimensional state space. The higher dimensional space can always be redefined in a one dimensional form. Thus every combination of values of x and y can be listed and the ith entry in the list can be called state i. Digital computers make this sort of conversion internally.

There is, however, the computational limitation mentioned in Chapter 1, that the number of states per stage should not exceed the order of 1 000. This restriction relates to states listed in one dimensional form. If the states are listed in multidimensional form with m dimensions and N states per dimension the limitation is that N^m must not exceed the order of 1 000. This restricts the value iteration method to problems with very few dimensions of state space.

Returning to the allocation problem, action k corresponds to the production of k items of the current product. This is a minor variation on the formulation used in the one resource problem. The return $r(n, k)$ is the income resulting from the production of k items of product n. The optimal value of state (n, x, y), denoted $f(n, x, y)$ is the total income when the system starts in state (n, x, y) and an optimal plan is followed. This specification is summarised in Table 2.19.

Table 2.19. TWO RESOURCE ALLOCATION PROBLEM: SPECIFICATION

Stage	Allocation of raw material to a product. n = number of products remaining	n
State	x tons of raw material X and y tons of raw material Y remain to be allocated and n products remain to be considered	(n, x, y) or (n, x_n, y_n)
Action	Produce k items of current product	k
Return	Income from k items of product n	$r(n, k)$
Optimal value of a state	Total income when the system starts in that state and an optimal plan is followed	$f(n, x_n, y_n)$

Variables are subscripted when this is convenient, thus x_n, y_n may be used to denote the quantities of raw material available at stage n. The quantity of raw material X required to produce k items of product n is denoted by $X(n, k)$, and the corresponding quantity of raw material Y by $Y(n, k)$. These are the quantities tabulated in Table 2.17. The key equations of the formulation are as follows.

Recurrence relation

$$f(n, x_n, y_n) = \underset{k \in K}{\text{Max}} \ [r(n, k)+f(n-1, x_{n-1}, y_{n-1})] \qquad (2.29)$$

Transition equations

$$\left. \begin{array}{l} x_{n-1} = x_n - X(n, k) \\ y_{n-1} = y_n - Y(n, k) \end{array} \right\} \qquad (2.30)$$

Return function $r(n, k)$ given by Table 2.18. Stage space, $0 \leqslant n \leqslant 3$. State space, all integer combinations of x and y subject to $0 \leqslant x \leqslant 4$, $0 \leqslant y \leqslant 4$. Action space, k a non-negative integer less than or equal to 3, subject to, $X(n, k) \leqslant x_n$, $Y(n, k) \leqslant y_n$. Terminal values are $f(0, x_0, y_0) = 0$.

2.9.5 Calculation

The calculation is shown in Table 2.20 where for compactness only the optimal actions and values appear. At stage 1 it is optimal to produce as much as possible within the action constraints. For subsequent stages the general procedure for evaluating the trial value of state (n, x_n, y_n) under action k is as follows.

1. Look up the quantities of raw material used, $X(n, k)$ and $Y(n, k)$ in Table 2.17.
2. Determine the successor state $(n-1, x_{n-1}, y_{n-1})$ using the transition equation. If x_{n-1} or y_{n-1} is negative the action constraint operates and optimization at this state is concluded.
3. Look up $r(n, k)$ in Table 2.18 and $f(n-1, x_{n-1}, y_{n-1})$ in Table 2.20.
4. Calculate the trial value $r(n, k)+f(n-1, x_{n-1}, y_{n-1})$.

For example consider state $(2, 3, 2)$. For actions $k = 0, 1$ the steps just described yield the following results. Actions $k = 2, 3$ are excluded

Table 2.20. TWO RESOURCE ALLOCATION PROBLEM:
CALCULATIONS

Stage	State		Action	Value
	x	y	(optimal)	(optimal)
0	all			0
all	x or y = 0			0
1	1	1	1	5
1	1	2	2	8
1	1	3	2	8
1	1	4	2	8
1	2	1	1	5
1	2	2	2	8
1	2	3	3	11
1	2	4	3	11
1	3	1	1	5
1	3	2	2	8
1	3	3	3	11
1	3	4	3	11
1	4	1	1	5
1	4	2	2	8
1	4	3	3	11
1	4	4	3	11
2	1	1	0	5
2	1	2	0	8
2	1	3	0	8
2	1	4	0	8
2	2	1	0	5
2	2	2	0	8
2	2	3	3	17
2	2	4	3	17
2	3	1	0	5
2	3	2	0	8
2	3	3	3	17
2	3	4	3	22
2	4	1	0	5
2	4	2	0	8
2	4	3	3	17
2	4	4	3	22
3	4	4	2	25

by the action space constraint.

$k = 0$	$k = 1$
$X(2, 0) = 0$	$X(2, 1) = 1$
$Y(2, 0) = 0$	$Y(2, 1) = 2$
$x_1 = 3-0 = 3$	$x_1 = 3-1 = 2$
$y_1 = 2-0 = 2$	$y_1 = 2-2 = 0$
$r(2, 0) = 0$	$r(2, 1) = 6$
$f(1, 3, 2) = 8$	$f(1, 2, 0) = 0$
$r(2, 0)+f(1, 3, 2) = 8$	$r(2, 1)+f(1, 2, 0) = 6$

The larger trial value is obtained under action $k = 0$ and the optimal value of state $(2, 3, 2)$ is $f(2, 3, 2) = 8$. This result appears in Table 2.20, line 28.

The optimal plan is picked out from Table 2.20 by the procedure of Figure 1.4, and is to make 2 items of product A, none of product B and 2 of product C. The optimal process is shown in Table 2.21.

Table 2.21. TWO RESOURCE ALLOCATION
PROBLEM: OPTIMAL PROCESS

Stage	State		Action	Value
	x	y		
3	4	4	2	25
2	2	2	0	8
1	2	2	2	8

2.10 LAGRANGE MULTIPLIER

A two resource allocation problem such as that just considered can be reduced to a one resource allocation problem by giving one of the resources a linear cost, referred to in general as a Lagrange multiplier. Suppose that raw material Y costs λ per unit. The return associated with producing and selling k units of product n is reduced by $\lambda Y(n, k)$. For any given numerical value of λ the problem of allocating the single resource X to products A, B, C, with the modified return function $(r(n, k) - \lambda Y(n, k))$ can be solved using the recurrence relation

$$f(n, x_n) = \underset{k \in K}{\text{Max}} \left[(r(n, k) - \lambda Y(n, k)) + f(n-1, x_{n-1}) \right]$$

$$f(0, x_0) \quad \text{given}$$

By trying several values of λ it may be possible to find one for which the available amount of the second resource is just used up; the solution then obtained is optimal for the original two resource problem. Unfortunately there is no guarantee that a value of λ will exist for which the second resource is exactly used up. Exercises Q2.11, Q2.12, Q2.13 illustrate these points. See also Nemhauser 1966.

The difficulty of the possible non existence of an optimising value of the Lagrange multiplier can be circumvented by finding the 'kth optimal policy' a technique detailed by Roberts 1964, which can also be useful in sensitivity analyses.

2.11 THE GENERAL ALLOCATION PROBLEM

The general allocation problem is

$$\left. \begin{array}{l} \text{maximise} \qquad \displaystyle\sum_{n=1}^{N} r(n, k) \\[2em] \text{subject to} \quad \displaystyle\sum_{n=1}^{N} X(n, s, k) \leqslant c(s), \qquad s = 1, 2, \ldots, m \end{array} \right\} \qquad (2.31)$$

In terms of 'products' and 'raw materials' this is equivalent to maximising the sum of the returns associated with producing N products, given that the return associated with producing k items of the nth product is $r(n, k)$. There are m raw materials and production of k items of product n consumes quantity $X(n, s, k)$ of the sth raw material. Total consumption of the sth raw material must not exceed $c(s)$.

The dynamic programming formulation follows the same lines as for the two resource problem. There is a dimension of state space for each raw material. Let $x_n(s)$ denote a general value of the quantity of the sth raw material remaining at stage n. A state is defined by the variables

$n, x_n(1), x_n(2), \ldots, x_n(m)$. The recurrence relation is

$$f(n, x_n(1), \ldots, x_n(m)) = \underset{k \in K}{\text{Max}} \; [r(n, k) + f(n-1, x_{n-1}(1), \ldots, x_{n-1}(m))] \qquad (2.32)$$

The linear allocation problem

$$\left. \begin{array}{ll} \text{maximise} & \displaystyle\sum_{n=1}^{N} r_n \cdot k_n \\[3mm] \text{subject to} & \displaystyle\sum_{n=1}^{N} X_{ns} \cdot k_n \leqslant c(s) \end{array} \right\} \qquad (2.33)$$

where r_n and X_{ns} are constants, is a special case of problem 2.31, above. Thus any linear programming problem can be formulated as a dynamic programme, although it is much more efficient to use linear programming for allocation problems when this is possible. Dynamic programming can be used to advantage in moderate sized nonlinear problems, particularly where variables are integer and there are state and action constraints.

EXERCISES

Q.2.1. A boatbuilder has orders for boats of a certain type to be delivered at the end of the months shown.

Boatbuilders orders

Month	No. of boats
Feb	1
Mar	2
Apr	5
May	3
June	2
July	1

Stock is zero at the beginning of February and is to be zero after the July delivery. If boats are built in a particular month there is an overhead cost of 4 units. Stock holding costs 1 unit per boat per complete month. Demands must be met. In what months should boats be built and in what quantities to minimise costs?

Notes

This problem is similar to the boatbuilding problem discussed earlier in this chapter but without the constraints on the stock level and build quantity. This makes it possible to use an efficient progressive formulation for which the following is an outline.

State	Situation where stock level is zero and n months remain	n
Action	Build sufficient boats to supply demand for k months	k
Return	Cost of building and storage associated with producing all boats for months n to $n-k+1$ in month n.	$r(n, k)$

The recurrence relation is

$$f(n) = \operatorname*{Min}_{0<k\leqslant n} [r(n, k)+f(n-k)] \qquad (2.34)$$

$$f(0) = 0.$$

The solution in terms of the number of months stock to built in each month is Feb 2, Apr 2, June 2; or Feb 2, Apr 1, May 3. Total cost 18 units.

Q.2.2. (a) An itinerent trader can transport a maximum of 140 ft³ of goods. He sells four different types of item, A, B, C, D. The volume per item is as follows $A = 10$ ft³, $B = 30$ ft³, $C = 40$ ft³, $D = 60$ ft³. The expected profit associated with carrying various levels of stocks of the items is as shown in Table 2.22.

Table 2.22. PROFIT LEVELS

	Type			
Stock level	*A*	*B*	*C*	*D*
1	1	10	14	22
2	2	18	28	40
3	3	26	38	50
4	4	30	44	—
5	5	32	—	—

How many of each item should be transported in order to maximise his expected profit and what will his expected profit then be?

Answer:

> 0 of *A*, 0 of *B*, 2 of *C*, 1 of *D*.
> Expected total profit £50.

(b) Making use of the calculations carried out for part (a) determine the trader's optimal load and profit if he has 190 ft³ of carrying capacity but must carry at least one of each item.

Answer:

> 2 of *A*, 1 of *B*, 2 of *C*, 1 of *D*.
> Expected total profit £62.

Q2.3. Fish farming problem

A fish farmer keeps fish in a single large tank. He has a contract to supply a fixed number of fish at the end of each month for the period January to May. The price he receives for the fish depends on the month and the size of the fish in the way shown in Table 2.23.

Table 2.23. SELLING PRICE, PENCE PER FISH

Month	Length of fish (in)				
	8	9	10	11	12
Jan	10	15	20	—	—
Feb	8	10	14	20	24
Mar	4	6	10	14	20
Apr	2	4	8	14	20
May	0	2	8	14	22

At the beginning of January all the fish are 8 in long. Subsequently they can be fed at rates 1, 2 or 3 in each month. Feeding at rate 1 maintains a fish at its current size, feeding at rate 2 increases its size by one inch in one month and feeding at rate 3 increases its size by two inches in one month. All the remaining fish must be fed at the same rate. The cost of feeding at the various rates depends on the size of the fish in the way shown in Table 2.24.

Table 2.24. FEEDING COST, PENCE PER FISH

Length (inches)	Rate		
	1	2	3
8	1	2	5
9	3	4	7
10	5	6	7
11	8	9	—
12	10	—	—

Determine the feeding policy which maximises the total return and the total return under that policy.

Answer: Feeding rates 1, 1, 2, 2, 3. Profit 24 pence per fish.

Q2.4. A plant overhaul has to be completed in 10 days. The overhaul has 3 stages, stripping, repairing and rebuilding, and these stages must be completed successively. The time taken to complete each stage is related to the cost of the resources which are used at that stage in the way shown in Table 2.25.

Table 2.25.

Time to complete in days	Cost of stage (£1 000s)		
	Stripping	Repairing	Rebuilding
2	18	11	20
4	17	8	15
6	8	7	9
8	6	5	8

Develop a general dynamic programming formulation for allocating the available time to the stages of the overhaul in such a way as to minimise costs, and use your formulation to obtain the optimal allocation with the given data.

Answer:

2 days stripping
2 days repairing
6 days rebuilding

Q2.5. (a) A man must each year allocate his current capital between gross income and investment.

Money taken as gross income yields nett income equal to the square root of the gross. Money invested yields interest at 10% per annum. Given an initial sum of £10 000 how should the man divide it so as to maximise his total nett income over a period of three years.

Answer: First year take as gross income £3 020; second year, £3 658; third year £4 422. Total nett income £181·7. (Problems of this type are discussed extensively by O.L.R. Jacobs 1967.)

Q2.5. (b) The man decides that instead of maximising his total nett income he wishes to maximise his minimum annual nett income. How should he then divide his money?

Answer: Take as gross income £3 656 in each year. Nett income is £60·46 in each year.

Q2.6. A reactor in a process plant is shut down annually and either overhauled or replaced. The cost of overhaul is related to the age of the reactor as shown in Table 2.26.

Table 2.26.

Age, years	1	2	3	4	5
Cost of overhaul thousands of £	1	4	9	15	21

The cost of a new reactors is £20 000. The life of the whole plant is 12 years starting from new. Determine an optimal policy for the replacement of the reactor over the 12 year period, and the total cost under this policy excluding the cost of the original reactor.

Answer: Replace at the end of the third, sixth and ninth years. Total cost £80 000.

Q2.7. A company has to transport N passengers and can use aircraft of types 1, 2, 3. The cost of using $x \geqslant 1$ aircraft of type 1 is $a_1 + b_1 x$ and the number of seats in an aircraft of type 1 is c_1, where a_1, b_1 and c_1 are known constants. Similar expressions hold for aircraft types 2 and 3. Develop a general dynamic programming formulation to determine how many aircraft of each type should be provided to transport the passengers at minimum cost. Use your formulation to solve the problem when $N = 563$ and the remaining constants are as shown in Table 2.27.

Table 2.27.

Aircraft type, n	Constants a_n	b_n	c_n
1	10	3	200
2	8	3	170
3	5	2	80

Answer: Three of type 1, zero of type 2, zero of type 3.

Q2.8. A company can buy t tons of a raw material for £ $c(t)$. It can use the raw material to make products A, B, C in various amounts. The return resulting when k tons of raw material are allocated to the production of A is $r(A, k)$ and similar relationships apply for B and C. Use dynamic programming to formulate a method for determining how much raw material the company should buy and how much it

Table 2.28. COST OF RAW MATERIAL

t tons	2	4	6	8
£ $c(t)$	15	28	40	50

should allocate it to the various products. Using your formulation solve this problem with the data shown in Tables 2.28 and 2.29. Note that the variables k and t can only take the values which appear in the tables, or zero.

Table 2.29. RETURNS FROM RAW MATERIAL USAGE

| Allocation a tons | | Returns (£) | |
	$r(A, k)$	$r(B, k)$	$r(C, k)$
1	9	6	8
2	18	12	10
3	23	21	11

Answer: Buy 6 tons and allocate 2 tons to Product A, 3 tons to Product B and 1 ton to Product C. Total nett return is £7.

Q2.9. Nonlinear transportation problem.

A company has two depots D_1, D_2 from which it supplies four customers C_1, C_2, C_3, C_4. At depot D_1 there are 5 items of a certain product and at D_2 there are 3 items of the same product. Customers C_1, C_2, C_3, C_4 require 1, 3, 2, 2 items respectively. Up to two items can be transported from any depot to any customer and the costs of transportation for carrying one and two items are as shown in Tables 2.30 and 2.31.

Table 2.30. TRANSPORTATION COSTS FOR ONE ITEM

From	To	C_1	C_2	C_3	C_4
D_1		1	4	5	6
D_2		3	2	7	4

Table 2.31. TRANSPORTATION COSTS FOR TWO ITEMS

From	To	C_1	C_2	C_3	C_4
D_1		2	6	8	7
D_2		5	4	10	9

Develop a dynamic programming formulation for determining the cheapest routing of the items from two sources to n destinations with nonlinear (and linear) costs and use it to solve the problem given. (See Hadley 1964.)

Answer: From D_1, 1 to C_1, 2 to C_2, 2 to C_3, remainder from D_2. Cost 26, there are the solutions.

Q2.10. A construction company has a work schedule which if no overtime is worked requires in a certain trade the following numbers of men in the months indicated. March 4, April 6, May 7, June 4, July 6, August 2. Three men are employed in February and it is planned to have three employed in September. The cost of hiring a man is £50 and the cost of firing a man is £80. It is not feasible to recruit more than three men in any month and under a union agreement not more than one third of the tradesmen can be fired in any month. The cost of having surplus manpower is £100 per man per month and the cost of having a manpower shortage (which must be made up in overtime) is £200 per man per month. Overtime cannot exceed 25% of normal time. Determine the hiring and firing policy which minimises costs. Assume that staff changes occur at the end of a month.

Answer: Hire, Feb 1, Mar 2, Apr 1, May −1, June 0, July −2, August −1. Total cost £920.

Q2.11. A factory can produce and sell goods of types *A*, *B* and *C* and earn income as shown in Table 2.32.

Table 2.32. INCOME FROM SALE OF
VARIOUS QUANTITIES OF GOODS

Qty Produced	A	B	C
1	5	6	10
2	9	12	18
3	12	17	25

The items require raw materials *X* and *Y* in the quantities shown in Table 2.33.

Table 2.33. RAW MATERIAL REQUIREMENTS

Qty Produced	A		B		C	
	X	Y	X	Y	X	Y
1	2	1	1	1	1	2
2	3	1	2	2	2	3
3	3	2	3	2	2	4

4 units of X and 4 units of Y are available.

How many items of each type should be produced to maximise total income?

Answer: 0 of A, 3 of B, 1 of C, total income 27 units.

Q2.12. Solve Q2.11 by the Lagrange multiplier method. Hint: Assume that raw material Y costs λ per unit. For example if $\lambda = 2$ the net return from producing one unit of product A is reduced from 5 to 3. Similarly the other returns are reduced so that the table of returns is as shown in Table 2.34.

Table 2.34. NET RETURNS WHEN $\lambda = 2$

Qty Produced	A	B	C
1	3	4	6
2	7	8	12
3	8	13	17

Solve the problem as a single resource allocation problem with the returns shown in Table 2.34. The solution is to produce 0 of A, 2 of B, 3 of C and this requires 6 units of raw material Y. Since only 4 units of Y are available we have priced it too cheaply. Try $\lambda = 5$.

Q2.13. A factory can produce and sell goods of types A, B and C and earn income as shown in Table 2.32. The items require raw materials X and Y in the quantities shown in Table 2.35.

Table 2.35. RAW MATERIAL REQUIREMENTS

Qty Produced	A		B		C	
	X	Y	X	Y	X	Y
1	2	1	1	1	1	2
2	3	1	2	2	2	3
3	3	1	3	2	2	4

How many items of each type should be produced to maximise total income? Can the problem be solved by the Lagrange multiplier method? (Try $\lambda = 4\frac{3}{4}$.)

Answer:

There is no value of λ for which 4 units of Y is used. The problem can be solved by the full two resource allocation method and is 0 of A, 3 of B, 1 of C, total income 27 units. See also the kth optimal policy method (Roberts 1964).

Q2.14. A company makes a chemical product by a reproductive process. The product is handled only in units of 1 000 gal. Inventory (stock plus work in progress) must always lie in the range 1 to 5 units. In any week one or more units of the chemical can be reproduced so that inventory is at most doubled, subject to the 5 unit maximum. Deliveries occur at the end of a week. The stock at the start of week 1 is 1 unit. Demand in the subsequent 5 week period is as shown in Table 2.36.

Table 2.36.

Week	Demand
1	2
2	3
3	1
4	4
5	5

Back orders are not met. Determine a production and delivery schedule which minimises the unsatisfied demand.

Answer: See Table 2.37

Table 2.37.

Week	Produce	Deliver
1	1	0
2	2	2
3	2	1
4	2	$\left.\begin{array}{c}2\\4\end{array}\right\}$ or $\left.\begin{array}{c}3\\3\end{array}\right\}$
5	2	

Total number of shortages is 6.

Q2.15. A company has the demands shown in Table 2.38 for a product over a four month period.

Table 2.38.

Month	Demand
Jan	3
Feb	2
Mar	1
Apr	3

Items can be made in the month they are required or made earlier and held in stock. Stock holding costs one unit per item per month and the maximum stock is four items. If items are produced in any month there is a set-up cost of 4 units. The production in any month cannot exceed four items. The stock at the beginning of January is two items and the value of stock at the end of April is one unit per item.

Demands must be met and only integer numbers of items can be produced. Determine a minimum cost production schedule.

Answer: Make; Jan 4, Feb 0, Mar 0, Apr 4.
  Total cost 13 units.

Q2.16. In an assembly shop, engines are built by adding components to a basic casting. The cost of adding a component depends on which other components have already been added. Develop a dynamic programming formulation to minimise the cost of assembly.

Q2.17. A metal is produced in sheets four feet square. The sheets are marked with a square grid of side one foot and origin at a corner of a sheet as shown.

The sheets can have defects whose location can be identified. The shading in the diagram indicates a defective area. The sheets are to be cut to provide square pieces of metal without defects. The return associated with a square of side w is $w^{5/2}$.

The cutting is done by two passes through a guillotine. All cuts are along grid lines. At the first pass the sheets are cut into rectangular

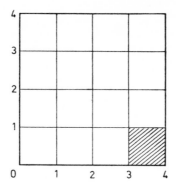

sections whose longer side is four feet. At the second pass the sections are fed individually and are cut parallel to their shorter sides.

Formulate an algorithm for computing a sequence of cuts which maximises the value of the pieces for any given pattern of defects and use your formulation to determine the optimum sequence of cuts for the sheet shown.

Notes: This is a simplified version of a problem described by Susan Hahn, Operations Research, 1968, 16, 1100–1114.

Answer: First pass; cut along vertical line $x = 1$.

Second pass; first section, cut along $y = 1, 2$ and 3
second section cut along $y = 1$.
Total value 19.29.

CHAPTER 3

Finite Stage Markov Programming

3.1 INTRODUCTION

The theory and applications of dynamic programming discussed in Chapters 1 and 2 were concerned with deterministic problems, where, under a given plan, a system moved with certainty from one state to another at each stage, generating a series of known returns. We now consider problems in which the transitions between states and the returns generated are governed by probabilistic laws. A process which involves a sequence of random variables is called a stochastic process (Bartlett, 1955, Feller). There is a range of levels of information which we may have about such a process (Wald, 1950, Raiffa and Schlaifer, 1961). At one extreme are processes where nothing is known about the distributions of the random variables; this is known as a situation of *uncertainty*. A second case is where we assume knowledge of the type of distributions of the random variables but rely on observation of realised values to refine our estimates of the parameters of these distributions. This leads to *adaptive control problems*. (Bellman, 1962, White, 1969, Nemhauser, 1966.) The type of situation which is considered in this chapter entails full knowledge of the distributions of the random variables. This is known as a situation of *risk*.

3.2 MARKOV PROCESSES

A stochastic process involves a sequence of random variables X_1, X_2, ..., X_n, ... The probability that the nth random variable, X_n, takes the value x_n may depend on the values taken by all the previous random variables, so that in general one considers the conditional probability,

$$\text{Prob}\,(X_n = x_n \mid X_1 = x_1, \ldots, X_{n-1} = x_{n-1}) \tag{3.1}$$

A model which assumes that the current variable depends on many previous outcomes is difficult to handle computationally. Adequate and computationally feasible models can often be established by considering the subclass of stochastic processes which are known as Markov processes, in honour of the Russian mathematician A. A. Markov (1856–1922) who was the first to study them. In a Markov process the probability that the random variable X_n takes the value x_n depends only on the immediately preceeding outcome x_{n-1}.

$$\begin{aligned}\text{Prob}\,(X_n = x_n \mid X_1 = x_1, \ldots, X_{n-1} = x_{n-1}) \\ = \text{Prob}\,(X_n = x_n \mid X_{n-1} = x_{n-1})\end{aligned} \tag{3.2}$$

To relate the general Markovian property stated in equation 3.2 to the processes studied in this chapter we link the two ideas in the following way. The serial problems considered in earlier chapters have involved a system which undergoes a process which proceeds by stages. When n stages remain the system can be in any of a number of states $(n, 1)$, $(n, 2)$, ..., (n, N). Consider the sequence of random variables ..., X_n, X_{n-1}, ... Let the domain of X_n be the set $\{1, 2, \ldots, N\}$. The conditional probability

$$\text{Prob}\,(X_{n-1} = j \mid X_n = i)$$

is the probability that the system moves to state $(n-1, j)$ given that it was at state (n, i) at stage n. This is written more compactly as

$$p(n, i, j) = \text{Prob}\,(X_{n-1} = j \mid X_n = i) \tag{3.3}$$

The *transition probabilities* $p(n, i, j)$ have the properties

$$0 \leqslant p(n, i, j) \leqslant 1 \tag{3.4}$$
$$\sum_{j=1}^{N} p(n, i, j) = 1$$

If the transition probabilities are stationary in n they can be written $p(i, j)$. A process with stationary transition probabilities is a stationary (time homogeneous) process. The limiting properties of infinite stage, stationary Markov processes are considered in Chapter 4. In this chapter finite stage processes are considered and are assumed to be nonstationary, the stationary case being a simple specialisation.

3.2.1 Marketing Example

The following is an example of a Markov process. A company which markets fashion goods has a product line which at the start of any year can be regarded as successful (state variable $i = 1$) or unsuccessful (state variable $i = 2$). The company considers its prospects over a finite planning horizon. When n years remain in the planning period the system is in general in state (n, i). The probability of transition from state (n, i) to state $(n-1, j)$ is $p(n, i, j)$.

In particular assume a planning period of two years and that the product is currently successful, state $(2, 1)$. The company makes the following assessment regarding the performance of the product in the planning period. The probability that the product will still be successful at stage 1 (that is, when one year remains) is 0·8. The probability that it will then be unsuccessful is therefore 0·2. If the product is successful at stage 1 the probability that it will be successful at stage 0 (that is, on reaching the planning horizon) is 0·9, and the probability that it will be unsuccessful at stage 0 is 0·1. These latter probabilities are conditional; they apply given that state $(1, 1)$ is reached. If the product is unsuccessful at stage 1 the probability that it will be successful at stage 0 is 0·4. The transition probabilities can be summarised in matrix form as follows

$$\left. \begin{aligned} \mathbf{P}(2) = [p(2, i, j)] &= \begin{bmatrix} 0·8 & 0·2 \end{bmatrix} \\ \mathbf{P}(1) = [p(1, i, j)] &= \begin{bmatrix} 0·9 & 0·1 \\ 0·6 & 0·4 \end{bmatrix} \end{aligned} \right\} \qquad (3.5)$$

The process is represented by Figure 3.1, where the circles indicate the states and the lines represent the various possible transitions. The numbers against the lines are the corresponding transition probabilities.

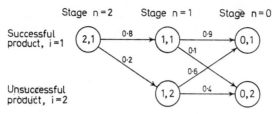

Figure 3.1. Marketing example: Markov process

3.3 MARKOV PROCESSES WITH RETURNS

Suppose that if a system goes from state (n, i) to state $(n-1, j)$ a return $c(n, i, j)$ is generated. This return is called a *transition return* because it is associated with a particular transition between states. Let the returns be additive. Define the value of a state as the average total return which is generated when that state is the starting state and denote the value of state (n, i) by $f(n, i)$. Note that we are considering a single process only and that the question of optimality does not arise yet. The value of state (n, i) is related to the values of states $(n-1, j)$, $j = 1, 2, \ldots, N$, by

$$f(n, i) = \sum_{j=1}^{N} p(n, i, j)(c(n, i, j) + f(n-1, j)) \qquad (3.6)$$

Define the *stage return* $r(n, i)$, associated with state (n, i) as the average return generated at the current stage, given that the system starts the stage in state (n, i). Then $r(n, i)$ is given by

$$r(n, i) = \sum_{j=1}^{N} p(n, i, j) \, c(n, i, j) \qquad (3.7)$$

Using this definition, equation 3.6 can be written

$$f(n, i) = r(n, i) + \sum_{j=1}^{N} p(n, i, j) \, f(n-1, j) \qquad (3.8)$$

To illustrate this we add a return structure to the marketing example. The product is successful initially. Suppose that if the product is still successful at stage 1 there is a profit of 1 unit. This is the transition return associated with transition from state (2, 1) to state (1, 1) and is denoted $c(2, 1, 1) = 1$. If the product has become unsuccessful by stage 1 there is a loss of two units. This is denoted $c(2, 1, 2) = -2$. These and the other returns which follow would have to be assessed by the company. The returns at stage 1 are as follows. If the product is successful at the start of the year and remains successful at the end there is a profit of 2 units, $c(1, 1, 1) = 2$. If the product becomes unsuccessful, however, there is a loss of one unit, $c(1, 1, 2) = -2$. If the product is unsuccessful at the start of the year but successful by the end there is a profit of 1 unit, $c(1, 2, 1) = 1$, but if the product is still unsuccessful at the end there is a loss of three units, $c(1, 2, 2) = -3$. The transition returns can be summarised in matrix form

$$\mathbf{C}(2) = [c(2, i, j)] = \begin{bmatrix} 1 & -2 \end{bmatrix}$$

$$\mathbf{C}(1) = [c(1, i, j)] = \begin{bmatrix} 2 & -1 \\ 1 & -3 \end{bmatrix} \tag{3.9}$$

The transition returns are illustrated in Figure 3.2, which is similar to Figure 3.1 but with the numbers against each arc corresponding to the transition returns.

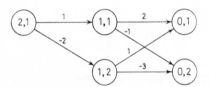

Figure 3.2. Marketing example: transition returns

The stage returns can be calculated using equation 3.7 and are

$$\left. \begin{array}{rcr} r(2, 1) = & 0{\cdot}4 \\ r(1, 1) = & 1{\cdot}7 \\ r(1, 2) = & -0{\cdot}6 \end{array} \right\} \tag{3.10}$$

The values of the states can be calculated using the recurrence relation 3.8. We must give values to the states $(0, 1)$ and $(0, 2)$ in order to initiate the recurrence. We put $f(0, 1) = 1$, $f(0, 2) = 0$, reflecting an assessed value advantage of one unit in favour of having a successful product when the planning horizon is reached. The values of the various states are then as shown in Table 3.1.

Table 3.1. MARKETING EXAMPLE: STATE VALUES

Stage	State	Value
0	1	1
0	2	0
1	1	2·6
1	2	0
2	1	2·48

3.4 MARKOV DECISION PROBLEMS

Suppose that in state (n, i) a set of actions K is available. Under a particular action k the probability of transition to state $(n-1, j)$ is $p(n, i, j, k)$. There is a corresponding transition return $c(n, i, j, k)$ associated with transition from state (n, i) to state $(n-1, j)$ under action k, and a stage return $r(n, i, k)$ associated with state (n, i) and action k. The stage return is given by

$$r(n, i, k) = \sum_{j=1}^{N} p(n, i, j, k) \, c(n, i, j, k) \qquad (3.11)$$

In the marketing example, suppose that the company has the following actions available in each state,

Action 1 = Advertise
Action 2 = Do not advertise.

The data already given refers to the situation where the company always advertises. The effect of not advertising is to reduce the probability that the product will continue or become successful, but at the same time to reduce costs and hence to increase returns. The data for the marketing example for both actions at all states is presented in Table 3.2.

With the inclusion of alternative actions we have a Markov decision problem. A set of actions, one for each state, constitutes a plan which defines a particular Markov process. The aim is to find the

7

Table 3.2. MARKETING EXAMPLE: DATA

State (n, i)	Action k	Transition probabilities		Transitions returns		Stage returns $r(n, i, k)$
		$p(n, i, 1, k)$	$p(n, i, 2, k)$	$c(n, i, 1, k)$	$c(n, i, 2, k)$	
2,1	1	0·8	0·2	1	−2	0·4
	2	0·6	0·4	3	1	2·2
1,1	1	0·9	0·1	2	−1	1·7
	2	0·7	0·3	4	1	3·1
1,2	1	0·6	0·4	1	−3	−0·6
	2	0·2	0·8	2	−1	−0·4

plan (and hence the process) which optimises the value of the initial state or states. In terms of the marketing example, we require the promotional plan with the maximum mean total return. The formulation and solution of Markov decision problems is known as Markov programming.

Finite stage Markov decision problems are solved by a probabilistic version of the value iteration algorithm. Let $f(n, i)$ be the optimal value of state (n, i), that is the mean total return when the system starts in state (n, i) and an optimal plan is followed. The following recurrence relation holds

$$f(n, i) = \underset{k \in K}{\text{Max}} \left[r(n, i, k) + \sum_{j=1}^{N} p(n, i, j, k) f(n - 1, j) \right] \quad (3.12)$$

The definitions of the leading terms in stochastic dynamic programming closely parallel those for deterministic problems, with the addition of the concepts of transition probabilities and stage returns. The interpretation of the leading terms in the stochastic case is summarised in the context of the marketing example in Table 3.3.

3.4.1 Solution of the Marketing Example

The marketing example is solved by application of the recurrence relation 3.12 using the data of Table 3.2 and the terminal values indicated earlier, that is, $f(0, 1) = 1$, $f(0, 2) = 0$. The calculations are shown in Table 3.4. The calculation is now shown in detail for state $(1, 1)$.

Table 3.3.   MARKETING EXAMPLE: SPECIFICATION

Stage	A year. n = number of years remaining until the planning horizon is reached	n
State	Successful product ($i = 1$) or unsuccessful product ($i = 2$) and n stages remaining	(n, i)
Action	Advertise ($k = 1$) or do not advertise ($k = 2$)	k
Transition probability	Probability of transition from state (n, i) to state $(n-1, j)$ under action k.	$p(n, i, j, k)$
Transition return	Return associated with transition from state (n, i) to state $(n-1, j)$ under action k	$c(n, i, j, k)$
Stage return	Mean return associated with state (n, i) and action k	$r(n, i, k)$
Optimal value of a state	Mean total return when the system starts in that state and an optimal plan is used	$f(n, i)$

The recurrence relation for state $(1, 1)$ is

$$f(1, 1) = \underset{k \,\in\, \{1,\, 2\}}{\text{Max}} \left[r(1, 1, k) + \sum_{j=1}^{2} p(1, 1, j, k) f(0, j) \right] \qquad (3.13)$$

The trial value of the test quantity under action 1 is

$$\left. \begin{aligned} & r(1, 1, 1) + \sum_{j=1}^{2} p(1, 1, j, 1) f(0, j) \\ & = 1 \cdot 7 + 0 \cdot 9 \times 1 + 0 \cdot 1 \times 0 = 2 \cdot 6 \end{aligned} \right\} \qquad (3.14)$$

For action $k = 2$

$$\left. \begin{aligned} & r(1, 1, 2) + \sum_{j=1}^{2} p(1, 1, j, 2) f(0, j) \\ & = 3 \cdot 1 + 0 \cdot 7 \times 1 + 0 \cdot 3 \times 0 = 3 \cdot 8 \end{aligned} \right\} \qquad (3.15)$$

Action $k = 2$ is optimal and $f(1,1) = 3 \cdot 8$. A similar analysis of the remaining states yields the other results shown in Table 3.4, where the optimal actions and values are italicised. The solution requires that we give an optimal action for every state, since it is possible to reach every state under the optimal plan. Thus none of the optimal actions in Table 3.4 is redundant, unlike in the deterministic case.

7*

Table 3.4. MARKETING EXAMPLE: CALCULATIONS

Stage	State	Action	Trial Value
0	1		*1*
0	2		*0*
1	1	1	$1{\cdot}7+0{\cdot}9\times1{\cdot}0+0{\cdot}1\times0 = 2{\cdot}6$
1	1	2	$3{\cdot}1+0{\cdot}7\times1{\cdot}0+0{\cdot}3\times0 = 3{\cdot}8$
1	2	*1*	$-0{\cdot}6+0{\cdot}6\times1{\cdot}0+0{\cdot}4\times0 = 0$
1	2	2	$-0{\cdot}4+0{\cdot}2\times1{\cdot}0+0{\cdot}8\times0 = -0{\cdot}2$
2	1	1	$0{\cdot}4+0{\cdot}8\times3{\cdot}8+0{\cdot}2\times0 = 3{\cdot}44$
2	1	2	$2{\cdot}2+0{\cdot}6\times3{\cdot}8+0{\cdot}4\times0 = 4{\cdot}48$

However, in any particular realisation of the optimal process only the actions for the states realised will be used. The optimal plan is to advertise in state (1, 2) but not to advertise in states (2, 1) and (1, 1). The mean total return is given by $f(2, 1) = 4{\cdot}48$ and is higher than that previously calculated for the case where only action 1 was available in each state.

3.5 COMPUTATIONAL LOAD

Consider a Markov decision problem with n stages, N states per stage and k actions per state. Solution by value iteration involves the following number of steps. Evaluation of the test quantity for given stage, state and action requires one addition for the stage return plus N multiplications and N additions for the summation term. k trial values must be calculated at each of nN states. Hence the total number of computational steps is

$$knN(2N+1) \tag{3.16}$$

or approximately $2N^2nk$. For the deterministic case the comparable result was $2Nnk$ (expression 1·34), showing an increase by a factor N in the probabilistic case.

3.6 DISCOUNTING

The use of discounting in deterministic investment decision problems was discussed in Chapter I. The same technique applies to Markov decision problems. Consider a system with general state (n, i), at which action k gives rise to a stage return $r(n, i, k)$ and to transition to state $(n-1, j)$ with probability $p(n, i, j, k)$, $j = 1, \ldots, N$. If future returns are discounted so that £1 received one stage hence has present value £b, then the optimal value $f(n, i)$ of state (n, i) obeys the recurrence relation

$$f(n, i) = \max_k [r(n, i, k) + b \sum_{j=1}^{N} p(n, i, j, k) f(n-1, j)] \quad (3.17)$$

Discounting is equivalent to the assumption that there is probability $(1-b)$ that the system will make a transition to an artificial state which has value zero. It can therefore be used to reflect uncertainty regarding the future viability of the system, as well as the cost of money.

3.6.1 Discounted Marketing Example

As an illustration of a finite stage Markov decision problem with discounting we solve the marketing example for which the data is given in Table 3.2, using a discount factor $b = 0.4$. This reflects a somewhat pessimistic view of the market in fashion goods.

The recurrence relation 3.17 applies. The terminal values $f(0, 1) = 1$, $f(0, 2) = 0$ are used as before. The optimisation at state $(1, 1)$ is as follows. The recurrence relation is

$$f(1, 1) = \max_{k \in \{1, 2\}} [r(1, 1, k) + 0.4 \sum_{j=1}^{2} p(1, 1, j, k) f(0, j)] \quad (3.18)$$

The trial values of the test quantity are

$$\left. \begin{array}{ll} k = 1: & 1.7 + 0.4(0.9 \times 1.0 + 0.1 \times 0) = 2.06 \\ k = 2: & 3.1 + 0.4(0.7 \times 1.0 + 0.3 \times 0) = 3.38 \end{array} \right\} \quad (3.19)$$

Action 2 is optimal and $f(1, 1) = 3.38$. The full calculations are shown in Table 3.5. The optimal actions and values are italicised, and the optimal plan is not to advertise at all.

Table 3.5. DISCOUNTED MARKETING PROBLEM: CALCULATIONS AND SOLUTION

Stage	State	Action	Trial Value
0	1		*1*
0	2		*0*
1	1	1	$1 \cdot 7 + 0 \cdot 4(0 \cdot 9 \times 1 \cdot 0 + 0 \cdot 1 \times 0) = 2 \cdot 06$
1	1	2	$3 \cdot 1 + 0 \cdot 4(0 \cdot 7 \times 1 \cdot 0 + 0 \cdot 3 \times 0) = 3 \cdot 38$
1	2	1	$-0 \cdot 6 + 0 \cdot 4(0 \cdot 6 \times 1 \cdot 0 + 0 \cdot 4 \times 0) = -0 \cdot 36$
1	2	2	$-0 \cdot 4 + 0 \cdot 4(0 \cdot 2 \times 1 \cdot 0 + 0 \cdot 8 \times 0) = -0 \cdot 32$
2	1	1	$0 \cdot 4 + 0 \cdot 4(0 \cdot 8 \times 3 \cdot 38 + 0 \cdot 2 \times -0 \cdot 32) = 1 \cdot 456$
2	1	2	$2 \cdot 2 + 0 \cdot 4(0 \cdot 6 \times 3 \cdot 38 + 0 \cdot 4 \times -0 \cdot 32) = 2 \cdot 96$

3.7 STOCHASTIC INVENTORY MODELS

Inventory problems are concerned with rules governing the ordering of stock. They may relate to orders placed on an outside supplier or to the production of goods within the firm. The boatbuilder's problem, discussed in Chapter 2, is a deterministic example of the latter type. Inventory problems invariably involve a sequence of decisions in time and are well suited to analysis by dynamic programming.

In general there is demand for a certain product which is known or forecast over a finite or infinite planning period. Information about the demand may be deterministic, probabilistic with known distribution or generated by a system of adaptive forecasting. We consider here the case where demand is probabilistic with known distribution. The demand is normally met from stock. If there is a shortage (no stock), back demand may either accumulate or be lost. Inventory is replenished from time to time by placing orders on an external or an internal supplier. Supply may be available immediately or after a certain 'lead time'. Costs are associated with such factors as stockholding, shortages and the replenishment of stock. The aim is usually to find minimum cost or maximal return policies.

For some inventory problems quite simple rules suffice. Examples are the economic batch quantity formulae and the newsboy's formula described by, for example, Makower and Williamson 1967. Often the optimal ordering policy has the following form: If current stock exceeds level s do not order, otherwise order sufficient to bring stock up to level S. This is called an s, S policy. Conditions under which s, S policies are optimal are discussed by H. E. Scarf 1960. Certain models have been analysed in considerable depth. Beckmann 1968 gives an extensive analysis of several standard models using dynamic programming.

3.7.1 Stochastic Inventory Example

A retailer can place orders for gas cylinders at the start of a week and receive delivery at the start of the following week. If an order is placed there is a £5 delivery charge plus a charge of £10 per cylinder. The cylinders sell for £18 each. Storage costs are reckoned at £1 per cylinder per complete week. The demand for cylinders in week n is a random variable which takes the value x with probability $\phi(n, x)$, as shown in Table 3.6. Demand which is not immediately satisfied is lost. Cylinders remaining at the planning horizon, six weeks hence, are valued at £8 each. Determine an optimal ordering plan.

Table 3.6. STOCHASTIC INVENTORY
EXAMPLE: DEMAND PROBABILITIES $\phi (n, x)$

Demand x	Week n					
	6	5	4	3	2	1
0	0·3	0·5	0·3	0·1	0·1	0·5
1	0·3	0·5	0·5	0·2	0·2	0·4
2	0·4	0·0	0·2	0·4	0·3	0·1
3	0·0	0·0	0·0	0·3	0·4	0·0

3.7.2 Formulation

A week is a *stage* and n denotes the number of weeks remaining until the planning horizon is reached. *State* (n, i) corresponds to having stock level i at the start of week n. *Action k* corresponds to ordering k cylinders in the current state. The *stage return* $r(n, i, k)$ is the mean total of income minus expenditure in week n given stock i and order quantity k. The *probability of transition* from state (n, i) to state $(n-1, j)$

under action k is denoted $p(n, i, j, k)$. This corresponds to the probability that j cylinders will be in stock at the start of next week given that the current stock is i and k cylinders are ordered. The *optimal value* of state (n, i) is denoted $f(n, i)$ and is the mean total nett return when an optimal plan is followed from state (n, i). This formulation is summarised in Table 3.7.

Table 3.7. STOCHASTIC INVENTORY EXAMPLE: SUMMARY OF FORMULATION

Stage	A week	n
State	Stock level i at start of week n	(n, i)
Action	Order k cylinders	k
Stage return	Mean nett return in current week	$r(n, i, k)$
Transition probability	Probability that the stock level at the start of week $n-1$ is j, given that the stock level is i at the start of week n and that the order quantity in week n is k	$p(n, i, j, k)$
Optimal value of a state	Mean total nett return when the system starts in that state and an optimal plan is followed	$f(n, i)$

Since back demand is lost the minimum inventory position is $i = 0$. For computational purposes we introduce an arbitrary maximum stock level, N, set sufficiently high so as not to affect the optimal plan. The recurrence relation is then

$$f(n, i) = \underset{k}{\text{Max}} \left[r(n, i, k) + \sum_{j=0}^{N} p(n, i, j, k) f(n-1, j) \right] \quad (3.20)$$

The terminal values are given by

$$f(0, i) = 8i \quad (3.21)$$

The ranges of the variables are

$$\left. \begin{array}{l} 0 \leqslant n \leqslant 6 \\ 0 \leqslant i \leqslant N \\ 0 \leqslant k \leqslant N - i \end{array} \right\} \quad (3.22)$$

3.7.3 Calculation

The stage returns are calculated as follows. Let X denote the demand in the current week, a random variable. In the case where no orders are placed the state return is given by the mean income from sales minus the storage cost for unsold items.

$$\left. \begin{array}{ll} r(n,\ i,\ 0) = 18 \sum_{x=1}^{i} \text{Prob}\ (X \geqslant x) - \sum_{x=1}^{i} \text{Prob}\ (X < x), & i > 0 \\ r(n,\ 0,\ 0) = 0 & i = 0 \end{array} \right\} \quad (3.23)$$

If the order quantity is $k > 0$ the stage return is similar to equation 3.23 but reduced by the cost of the k items,

$$r(n, i, k) = r(n, i, 0) - (5 + 10k) \qquad k > 0$$

The transition probabilities are calculated as follows. In the case where no orders are placed and the stock level at the next stage is between i and 0 we have

$$p(n, i, j, 0) = \text{Prob}\ (X = (i - j)) \qquad i \geqslant j > 0$$

When the stock level at the next stage is 0 we have

$$p(n, i, 0, 0) = \text{Prob}\ (X \geqslant i)$$

and for $j > i$

$$p(n, i, j, 0) = 0 \qquad i < j$$

If the order quantity is $k > 0$ the transition probabilities are given by

$$p(n, i, j, k) = p(n, i, j - k, 0)$$

To illustrate the computational procedure consider state $(1, 1)$. If no cylinders are ordered the stage return is,

$$r(1, 1, 0) = 18\ \text{Prob}\ (X \geqslant 1) - \text{Prob}\ (X < 1)$$

Using the data of Table 3.6 we get

$$r(1, 1, 0) = 18 \times 0 \cdot 5 - 0 \cdot 5 = 8 \cdot 5$$

If k cylinders are ordered the stage return is

$$r(1, 1, k) = r(1, 1, 0) - (5 + 10k)$$
$$= 3.5 - 10k \qquad k > 0$$

If no cylinders are ordered the transition probabilities are

$$p(1, 1, 1, 0) = \text{Prob}(X = 0) = 0.5$$
$$p(1, 1, 0, 0) = \text{Prob}(X \geqslant 1) = 0.5$$

If k cylinders are ordered the transition probabilities are,

$$p(1, 1, k, k) = 0.5$$
$$p(1, 1, k+1, k) = 0.5$$

Using the terminal values, equation 3.21, and the recurrence relation equation 3.20, we get the optimal value of state $(1, 1)$ as follows

$$f(1, 1) = \underset{k}{\text{Max}} \left[\begin{array}{cc} k = 0 & k > 0 \\ 8.5 + 0.5 \times 0 + 0.5 \times 8 : 3.5 - 10k + 0.5(8k) \\ & + 0.5(8(k+1)) \end{array} \right]$$

Table 3.8. STOCHASTIC INVENTORY EXAMPLE: OPTIMAL PLAN

Stage (week) n	State (stock) i	Action (order qty.) k
1	all i	0
2	all i	0
3	0	3
3	1	3
3	2	3
3	3	2
3	4 or more	0
4	0	5
4	1	5
4	2	4
4	3 or more	0
5	0	5
5	1 or more	0
6	0	2
6	1 or more	0

Simplification yields

$$k = 0 \quad k > 0$$
$$f(1, 1) = \underset{k}{\text{Max}} \, [12{\cdot}5 \; : \; 7{\cdot}5 - 2k]$$

The maximum occurs when $k = 0$. Thus no cylinders should be ordered and

$$f(1, 1) = 12{\cdot}5$$

The full calculation is rather laborious and so it has been carried out by computer ,see Appendix. The optimal plan is shown is Table 3.8. The mean total return from state $(6, 0)$ under the optimal plan is £23·65.

3.8 PROGRESSIVE PROBLEMS

In a progressive problem a system has a finite set, U, of possible states which can be sequentially numbered, 0, 1, 2, ..., N, in such a way that transitions are only possible from states i to j where $j < i$. In a progressive Markov decision problem the probability of transition from state i to state j under action k is $p(i, j, k)$ and the associated transition return is $c(i, j, k)$. The stage return associated with state i and action k is $r(i, k)$, given by

$$r(i, k) = \sum_{j=0}^{i-1} p(i, j, k) c(i, j, k) \tag{3.24}$$

The general recurrence relation for the solution of such a problem by value iteration is

$$f(i) = \underset{k \in K_i}{\text{Max}} \, [r(i, k) + \sum_{j=0}^{i-1} p(i, j, k) f(j)] \tag{3.25}$$

where $f(i)$ is the optimal value of state i and K_i is the set of possible actions associated with state i.

The following example of a progressive Markov decision problem draws attention to a short cut method which can be used when the distribution of a random variable is independent of the action taken.

3.9 TELEVISION GUESSING GAME

In a television game a contestant opens a box containing a sum of money, £x, the amount of which is uniformly distributed in the range 0 to £10, and which he may accept or reject. If he accepts the game ends, otherwise he goes to the next box and repeats the procedure. This continues up to a maximum of four boxes. What amounts should the contestant accept or reject at each box in order to maximise his expected return?

3.9.1 Formulation

Let state i correspond to the situation where the contestant has just opened a box and i boxes remain including the current one. State 0 is the terminal state, that is the end of the game. Let action k in state i correspond to setting a control limit k, such that the money is accepted only if $x \geqslant k$. The transition probabilities $p(i, j, k)$ under action k are

$$\left. \begin{array}{l} p(i, i-1, k) = k/10 \\ p(i, 0, k) \quad = 1 - k/10 \end{array} \right\} \tag{3.26}$$

The stage return associated with state i and action k is $r(i, k)$ given by

$$r(i, k) = \int_{k}^{10} (x/10) \, \mathrm{d}x = 5 - k^2/20 \tag{3.27}$$

Let $f(i)$ be the mean return when an optimal plan is followed from state i. The recurrence relation is

$$f(i) = \underset{0 < k < 10}{\text{Max}} [r(i, k) + \sum_{j < 1} p(i, j, k) f(j)] \tag{3.28}$$

$$f(0) = 0$$

Using equations 3.26 and 3.27

$$f(i) = \underset{0 < k < 10}{\text{Max}} [5 - k^2/20 + k(f(i-1)/10)] \tag{3.29}$$

The optimal plan can be found by iteration of equation 3.29 from $f(0) = 0$ and is shown in Table 3.9. The optimal control limit at each stage can be found by differentiation of the test quantity with respect to k. However, the calculation can be accelerated by means of a short cut method which is now discussed.

3.9.2 Short Cut Method

In the television guessing game the short cut method can be derived most simply by differentiating the test quantity in equation 3.29 partially with respect to k. This gives

$$\partial[.]/\partial k = (f(i-1)-k)/10 \qquad (3.30)$$

The maximum occurs when $k = f(i-1)$, and we can therefore simplify equation 3.29 to the form

$$f(i) = 5+(f(i-1))^2/20 \qquad (3.31)$$

The result $k = f(i-1)$ indicates that the contestant should accept the money in the current box if it is more than the mean return associated with the suceeding box under an optimal plan. This result is independent of the particular numbers or the form of distribution chosen for the example, and similar results apply to other problems which have an appropriate structure. A more general description of this structure is now given, again using the television guessing game as an illustration.

Consider a system with states $i = 1, \ldots, N$, with associated actions k, stage returns $r(i, k)$ and transition probabilities $p(i, j, k)$. The value $f(i)$ of state i under an optimal plan is given by the recurrence relation

$$f(i) = \operatorname*{Max}_{k} \left[r(i, k) + \sum_{j=1}^{i-1} p(i, j, k)f(j) \right] \qquad (3.32)$$

Suppose that when the system arrives in state i the decision maker first observes the value, x, of a random variable and then takes an action which results in probabilistic transition to a successor state j. We can define a secondary set of states (i, x), corresponding to being in primary state i and finding that the random variable takes the value x. The values $f(i)$ and $f(i, x)$ of the primary states i and secondary states (i, x) are related by

$$f(i) = \sum_{\text{all } x} \phi(x) f(i, x) \tag{3.33}$$

where $\phi(x)$ is the probability that the random variable takes the value x.

In secondary state (i, x) let there be a set H of actions h. These differ from the actions associated with the primary states i and in practice are usually fewer in number. Let $r(i, x, h)$ be the stage return associated with state (i, x) and action h, and $p(i, x, j, h)$ be the probability of transition from secondary state (i, x) to primary state j under action h. Then

$$f(i, x) = \underset{h \in H}{\text{Max}} \left[r(i, x, h) + \sum_{j=1}^{i-1} p(i, x, j, h) f(j) \right] \tag{3.34}$$

Equations 3.33 and 3.34 imply

$$f(i) = \sum_{\text{all } x} \phi(x) \underset{h \in H}{\text{Max}} \left[r(i, x, h) + \sum_{j=1}^{i-1} p(i, x, j, h) f(j) \right] \tag{3.35}$$

Comparing equations 3.32 and 3.35 we see that the distribution of the random variable, over which the decision maker has no control, has been taken through the maximisation, which is now applied to the secondary states. The advantage of this lies in the fact that the returns and transition probabilities associated with the secondary states are often very simple. Thus in the television game the action set for state (i, x), which corresponds to being at box i and finding it to contain £x, is; accept the money $(h = 1)$ or reject the money $(h = 2)$. The corresponding stage returns and transition probabilities are

$$\begin{aligned} \text{Accept:} \ & r(i, x, 1) = x, \quad p(i, x, 0, 1) = 1 \\ \text{Reject:} \ & r(i, x, 2) = 0, \quad p(i, x, i-1, 2) = 1 \end{aligned} \tag{3.36}$$

From equations 3.35, 3.36 and the terminal value $f(0) = 0$ we get

$$f(i) = \sum_{\text{all } x} \phi(x) \left(\text{Max} \left[\overset{\text{Accept}}{x} : \overset{\text{Reject}}{f(i-1)} \right] \right) \tag{3.37}$$

Thus again we establish that the contestant should accept the money in box i if it is more than the optimal value of the adjacent box $i-1$. This again leads to equation 3.31 from which we can readily compute the optimal control limits and state values which are shown in Table 3.9.

Table 3.9. TELEVISION GUESSING GAME: SOLUTION

State i	Control Limit k	Value f(i)
1	0	5·0
2	5·0	6·25
3	6·25	6·90
4	6·90	7·38

Finally, Table 3.10 shows the specification of the television game in terms of both primary and secondary states.

Table 3.10. TELEVISION GUESSING GAME: SPECIFICATION

Primary State	No. of boxes remaining	i
Action	Set control limit	k
Stage return	Mean return at box i when control limit is k	$r(i, k)$
Transition probabilities	Probabilities of transition to successor primary states when control limit is k	$p(i, j, k)$
Optimal value of state i	Mean return from state i under an optimal plan	$f(i)$
Secondary State	Number x in box i	(i, x)
Action	Accept or reject £x	h
Stage return	Mean return at state (i, x) under action h	$r(i, x, h)$
Transition probabilities	Probabilities of transition to successor primary states under action h	$p(i, x, j, h)$
Optimal value of state (i, x)	Mean return when an optimal plan is followed from state (i, x)	$f(i, x)$

EXERCISES

Q3.1. A dealer places orders for a certain type of machine at the beginning of a month and receives delivery at the start of the next month. Owing to restrictions on credit and on the supply of machines the dealer cannot stock nor order more than one machine at a time. The cost of a machine is 8 units, the selling price is 10 units and there is a holding cost of one unit per machine per complete month. If the dealer places an order but finds subsequently that (owing to the maximum stock restriction) he is unable to take delivery he incurs a cost of one unit. At the beginning of January the dealer has one machine in stock. He assesses the probability of making a sale in each of the next four months, given that he has stock at the start of each month, as follows,

Jan	Feb	Mar	Apr
0·7	0·6	0·6	0·4

If he has no stock at the start of a month he cannot make a sale in that month. The dealer feels uncertain regarding sales more than four months ahead and decides to value stock held at the beginning of May at 7 units per item. Should the dealer place an order?

Answer: No.

Q3.2. A fishing vessel can fish in either a near fishing ground (action 1) or a far fishing ground (action 2). A round trip to the near fishing ground takes one week and results in a catch with probability 0·4 and in no catch with probability 0·6. A round trip to the far fishing ground takes two weeks and results in a catch with probability 0·9 and in no catch with probability 0·1. Decisions as to which ground to fish are made on a basis of a four week planning period after which the crew is paid off. The profit from the sale of a catch when n weeks remain in this period is $r(n)$ the current data being $r(3) = 10$ units, $r(2) = 9$, $r(1) = 10$, $r(0) = 9$. To which fishing grounds should the vessel sail in which weeks to maximise profits.

Answer:

Sail to near ground, then far ground, then near ground. Mean total return 16·6 units.

Q3.3. Reservoir Problem. A Water Authority catches and stores water in a reservoir and then distributes it for agricultural use. A probabilistic forecast of inflow to the reservoir has been made as shown in Table 3.11.

Table 3.11. PROBABILITIES ASSOCIATED WITH VARIOUS INFLOW LEVELS

Inflow (billion galls)	Month			
	Nov	Dec	Jan	Feb
0	0	0·6	0·9	0·3
1	0·4	0·4	0·1	0·3
2	0·6	0	0	0·3
3	0	0	0	0·1

Returns can be associated with various outflows in various months as shown in Table 3.12.

Table 3.12. RETURNS ASSOCIATED WITH VARIOUS OUTFLOW LEVELS, IN ARBITRARY PRICE UNITS

Outflow (billion galls)	Month			
	Nov	Dec	Jan	Feb
0	0	−1	−2	0
1	2	2	1	2
2	3	3	2	3
3 or more	3	3	3	3

The reservoir has a capacity of four billion gallons and inflow in excess of capacity goes to waste. Inflow in any month does not become available for use until the start of the following month. The Authority decides the outflow for each month at the start of that month. The value of water at the beginning of March is one price unit per billion gallons for the first two billion gallons and zero for extra quantities. The reservoir is always either full or three quarters full at the beginning of November. Determine the water supply plan which maximises the mean total return.

8

Answer:

Optimal outflows are as shown in Table 3.13.

Table 3.13.

Stock	Nov	Dec	Jan	Feb
1	—	1	1	1
2	—	1	1	2
3	1	1	2	2
4	2	1 or 2	3	2

Q3.4. Marriage problem.
A woman has five years in which to find a useful husband. In each year she encounters one prospective husband. The utility of prospective husbands is a random variable with the probability density function $\phi(x) = e^{-x}$. The woman assesses the utility of the next prospective husband and either accepts him or rejects him. If she accepts him the process terminates and she receives a return equal to the utility of the husband. If she rejects him she proceeds to the next candidate. What level of utility should she be prepared to accept in each year in order to maximise her expected return.

Answer:

Accept if utility exceeds 1·817 in year 1. 1·621 year 2, 1·368 year 3, 1 year 4, 0 year 5. Mean return is 1·980.

Q3.5. A man wishes to sell a house and he plans to auction it in four weeks time unless he can sell it privately before then. He assumes that the house will realise £5 000 at auction. In each of the intervening weeks there is probability 0·5 that a prospective purchaser will appear. Each purchaser makes a single offer the value of which is uniformly distributed over the range £5 000 to £6 000, and which the seller either accepts or rejects. Determine rules for accepting or rejecting offers so as to maximise the seller's expected return.

Answer:

Accept if offer exceeds £5 485 in week 1, £5 390 week 2, £5 250 week 3, £5 000 week 4. Mean return £5 550.

Q3.6. Replacement Problem.

A certain type of machine is inspected at annual intervals. The probability that a machine will be found to be satisfactory at its ith inspection (given that the inspection occurs) is $(1-h_i)$. If a machine is found to be unsatisfactory it must be either repaired at cost C or replaced by a new machine of the same type at cost A. Five year old machines must be replaced. Determine an optimal repair and replacement plan over a period of n years, given a new machine initially, and the following data. $A = 10, C = 5, h_1 = 0\cdot2, h_2 = 0\cdot1, h_3 = 0\cdot3, h_4 = 0\cdot6, n = 3$. Machines are worthless once the planning horizon is reached.

Q3.7. Solve problem Q3.6 with discount factor of $0\cdot9$.

Q3.8. A machine is inspected at annual intervals and the cost of the repair which is required to bring it up to some desired standard is estimated. A machine is repaired if the estimated cost is less than a limit known as the repair limit, otherwise it is replaced by a new machine of the same type. Machines can last for up to three years, a new one always being introduced if the end of the third year of life is reached. The machines are required to provide a service for a period of 5 years and initially we have a new machine. The repair cost at the end of the first year of life is uniformly distributed in the range 0–£100. The repair cost at the end of the second year of life is uniformly distributed over the range 0–£150. The cost of a new machine is £150. Determine optimum repair limits for the range of years and ages covered by the problem, which minimise the expected sum of repair plus replacement costs. Also determine the optimal deterministic replacement plan and compare its mean total cost with the mean total cost of the best repair limit plan.

(For discussion of this type of problem see Hastings 1970).

Q3.9. A company deals in a crop which is harvested annually. Stock is held in a warehouse which at the beginning of each trading year is either full or empty. In each year the harvest succeeds with probability 0.9 and fails with probability 0.1. The company can only buy the crop in years when the harvest succeeds. The following actions can be taken.

1. If the warehouse is full at the beginning of the year:
   (*a*) Buy and sell nothing, leaving warehouse full at start of next year.

(b) Sell the contents of the warehouse and purchase fresh stock (if this is possible) leaving the warehouse again full at the start of the next year provided the harvest succeeds and empty otherwise.

(c) Sell the contents and buy nothing.

2. If the warehouse is empty at the start of the year: the company can either:

(a) Buy nothing.

(b) Buy sufficient to fill the warehouse, provided the harvest succeeds.

The price of the commodity is largely influenced by government stockpiling. A forecast of the net price of buying and selling the commodity for the next five years is shown in Table 3.14.

Table 3.14.

Year	1	2	3	4	5
Buying price	10	13	18	14	11
Selling price	11	13	15	15	11

If the warehouse is kept full for a year with no stock movement there is a charge of one unit. Determine the optimum buying and selling plan for the five year period. Assume that the warehouse is full initially and that it is to be empty at the end of the planning period.

Answer: See Table 3.15.

Table 3.15.

Year	Warehouse	Action
1	Full	b
2	Full	b
2	Empty	b
3	Full	c
3	Empty	a
4	Full	c
4	Empty	a
5	Full	c
5	Empty	a

CHAPTER **4**

Infinite Stage Markov Programming

4.1 INTRODUCTION

The problems analysed in Chapters 1, 2 and 3 had finite planning horizons. Many systems in fact operate with no clear end point or break point in view, and we now consider the formulation and solution of sequential decision problems with remote planning horizons. As a preliminary we investigate the behaviour of a system undergoing a Markov process as the number of stages increases. Deterministic problems appear as a special case in which the transition probabilities are all 0 or 1.

A stochastic matrix, $\mathbf{P}$, is a matrix whose elements, $p(i,j)$, lie in the range $0 \leqslant p(i,j) \leqslant 1$, and whose rows sum to one. Consider a system with a finite number of discrete states, denoted by the state variable $i = 1, \ldots, N$. Let the conditional probability of transition from state i to state j be $p(i,j)$, independent of stage number. The transition probability matrix, $\mathbf{P}$, will be a square stochastic matrix of order N,

$$\mathbf{P} = [p(i,j)].$$

The following is an example of such a system.

4.1.1 Slotting Machine Example

A company operates a number of automatic machine tools which cut slots in the heads of screws. Each machine is inspected at regular intervals. A machine may be found to require resetting (state $i = 1$) or to be running satisfactorily ($i = 2$). Under a certain operating policy it is found that if a machine requires resetting at one inspection there is probability 0·4 that it will again require resetting at the next, and probability 0·6 that it will be satisfactory. If a machine is running at one inspection there is probability 0·2 that it will require resetting at the next and probability 0·8 that it will be running. These transition probabilities are summarised by the stochastic matrix,

$$\mathbf{P} = [p(i,j)] = \begin{bmatrix} 0·4 & 0·6 \\ 0·2 & 0·8 \end{bmatrix} \tag{4.1}$$

4.1.2 n stage Transition Probabilities

Suppose that we know that a system which has transition probability matrix $\mathbf{P}$ is currently in state i and that we wish to determine the probability that it will be in state j in n stages time, the stages being numbered forward in this instance. To analyse this problem in general terms let $\pi(n, i, j)$ denote the probability that the system is in state j at stage n given that it was in state i at stage 0. $\pi(n, i, j)$ is the n stage probability of transition from state i to state j. The n stage transition probabilities can be arranged as a square stochastic matrix $\mathbf{\Pi}(n)$, where

$$\mathbf{\Pi}(n) = [\pi(n, i, j)]$$

Now, $\mathbf{\Pi}(0) = I$, the identity matrix, and $\mathbf{\Pi}(1) = \mathbf{P}$. Also

$$\pi(n, i, j) = \sum_{k=1}^{N} \pi(n-1, i, k)\, p(k, j), \quad i = 1, \ldots, N; n = 1, 2, \ldots \tag{4.2}$$

The matrix equivalent of equation 4.2 is

$$\mathbf{\Pi}(n) = \mathbf{\Pi}(n-1)\mathbf{P} \qquad n = 1, 2, \ldots \tag{4.3}$$

Hence by induction

$$\mathbf{\Pi}(n) = \mathbf{P}^n \qquad n = 0, 1, 2, \ldots \tag{4.4}$$

Thus the probability that the system is in state j at stage n given that it was in state i initially is given by the ijth element of the matrix $\mathbf{P}^n$. The n stage transition probabilities for the slotting machine for $n = 0$ to 5 are shown in Table 4.1.

Table 4.1. SLOTTING MACHINE EXAMPLE: n STAGE TRANSITION PROBABILITIES

n	0	1	2	3
$\mathbf{P}^n$	$\begin{bmatrix} 1 & 0 \\ 0 & 1 \end{bmatrix}$	$\begin{bmatrix} 0.4 & 0.6 \\ 0.2 & 0.8 \end{bmatrix}$	$\begin{bmatrix} 0.28 & 0.72 \\ 0.24 & 0.76 \end{bmatrix}$	$\begin{bmatrix} 0.256 & 0.744 \\ 0.248 & 0.752 \end{bmatrix}$

n	4	5
$\mathbf{P}^n$	$\begin{bmatrix} 0.2512 & 0.7488 \\ 0.2496 & 0.7504 \end{bmatrix}$	$\begin{bmatrix} 0.25024 & 0.74976 \\ 0.24992 & 0.75008 \end{bmatrix}$

From Table 4.1 it looks as if $\mathbf{P}^n$ is approaching the limit

$$\begin{bmatrix} \frac{1}{4} & \frac{3}{4} \\ \frac{1}{4} & \frac{3}{4} \end{bmatrix}$$

and this is indeed the case. A system for which $\mathbf{P}^n$ tends to a limit as n tends to infinity is said to be *completely ergodic*. This is in contrast to a system which has some periodic states of which an example will shortly be given.

4.1.3 Classification of States

Consider the transition probability matrix

$$\mathbf{P} = \begin{bmatrix} \frac{1}{2} & \frac{1}{2} \\ 0 & 1 \end{bmatrix} \tag{4.5}$$

Suppose that the system is initially in state 1. At the first transition it may return to state 1 or it may enter state 2. If it enters state 2 it remains in that state indefinitely. A state to which return is uncertain is said to be *transient*, so that in the example just given state 1 is transient. States which are not transient are *recurrent*. Recurrent states can be divided into three types known as *absorbing, periodic* and *aperiodic* states. An absorbing state is one from which transitions to other states are impossible. In the example of equation 4.5 state 2 is absorbing. If i is an absorbing state then $p(i,j) = 0$ for $j \neq i$, and $p(i,i) = 1$.

A periodic state is one to which the system returns at regular intervals. The simplest example occurs with the transition probability matrix

$$\mathbf{P} = \begin{bmatrix} 0 & 1 \\ 1 & 0 \end{bmatrix}$$

for which

$$\mathbf{P}^n = \begin{bmatrix} 0 & 1 \\ 1 & 0 \end{bmatrix} \quad n \text{ odd}$$

$$\mathbf{P}^n = \begin{bmatrix} 1 & 0 \\ 0 & 1 \end{bmatrix} \quad n \text{ even}$$

$\mathbf{P}^n$ does not now approach a limit. The system oscillates regularly between the states and is said to be periodic. Even in this case, however, a form of limiting behaviour can be identified on a stagewise average basis.

Finally there are recurrent aperiodic states, also called *ergodic* states. A system which is in an ergodic state will return to that state at some future stage with probability 1, but unlike a periodic state we

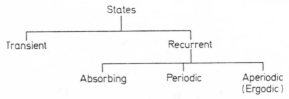

Figure 4.1. Classification of states by type

cannot say exactly when it will return. In the slotting machine example, equation 4.1, both states are ergodic.

The classification of states just described is summarised in Figure 4.1. A system can have states of some or all of these types.

4.1.4 Communication, Chain Structure and Irreducibility

If it is possible for a system which starts in state i to reach state j (not necessarily directly) then state i is said to *communicate* with state j. If state i communicates with state j and state j communicates with state i then states i and j *intercommunicate*. States can only intercommunicate with other states of the same type. Recurrent states can always be grouped into disjoint subsets within which the states intercommunicate and outside which they do not communicate. A system with a single set of recurrent states and possibly some transient states is said to be *unichain*. All the examples given so far are unichain. A system with several recurrent subsets is said to be *multichain*. A set of states all of which intercommunicate with one another is said to be *irreducible*.

4.2 ANALYSIS OF MARKOV PROCESSES

Associated with any stochastic matrix $\mathbf{P}$ are matrices $\mathbf{S}$ and $\mathbf{Y}$ which we refer to as the fixed and fundamental matrices respectively. They contain information about the behaviour of a system undergoing an infinite stage process with $\mathbf{P}$ as its transition probability matrix. They are now introduced, together with a matrix $\mathbf{E}$, known as the error matrix. For a fuller mathematical background to this material see Kemény and Snell 1960.

4.2.1 The Fixed Matrix

The ijth element of the matrix $\mathbf{P}^n$ gives the probability that a system is in state j at stage n given that it was in state i at stage 0. Suppose that a system which starts in state i is observed at a stage chosen at random during the next n stages. The probability that it will be found in state j is the ijth element of the matrix $\mathbf{S}(n)$, where,

$$\mathbf{S}(n) = (\mathbf{P} + \mathbf{P}^2 + \mathbf{P}^3 + \ldots + \mathbf{P}^n)/n \tag{4.6}$$

$\mathbf{S}(n)$ is a stochastic matrix. It can be shown that as n tends to infinity $\mathbf{S}(n)$ approaches a unique limit $\mathbf{S}$. We say that $\mathbf{P}$ is Cesaro summable to **S** where

$$\mathbf{S} = \operatorname*{Lim}_{n \to \infty} \mathbf{S}(n) \tag{4.7}$$

S is a stochastic matrix and is the *fixed matrix* associated with the given transition probability matrix **P**. The ijth element of **S** gives the limiting probability that the system will be in state j at a randomly chosen stage, given that it was in state i initially. It can further be shown that

$$\mathbf{SP} = \mathbf{PS} = \mathbf{S} \tag{4.8}$$

$$\mathbf{S} = \mathbf{S}^2 = \mathbf{S}^n \qquad n = 1, 2, 3, \dots \tag{4.9}$$

4.2.2 The Error Matrix

The error matrix **E** is defined by

$$\mathbf{E} = \mathbf{P} - \mathbf{S} \tag{4.10}$$

The rows of **E** sum to zero. Rearrangement of equation 4.10 gives

$$\mathbf{P} = \mathbf{S} + \mathbf{E} \tag{4.11}$$

Equations 4.8, 4.9 and 4.11 imply

$$\mathbf{SE} = \mathbf{ES} = 0 \tag{4.12}$$

Raise both sides of equation 4.11 to the nth power and use equations 4.9 and 4.12. This gives the important result

$$\mathbf{P}^n = \mathbf{S} + \mathbf{E}^n \qquad n = 1, 2, 3, \dots \tag{4.13}$$

Thus the ijth element of $\mathbf{E}^n$ gives the difference between the probability of finding the system in state j at stage n and the limiting probability if finding the system in state j at a randomly chosen stage, given in each case that the system started in state i. For completely ergodic systems $\mathbf{E}^n$ tends to zero and $\mathbf{P}^n$ tends to **S**. Equations 4.6, 4.7 and 4.13 imply that in general the limiting stagewise average of the error terms is zero. We say that **E** is Cesaro summable to zero

$$\operatorname*{Lim}_{n \to \infty} \left(n^{-1} \sum_{r=1}^{\infty} \mathbf{E}^r \right) = 0 \tag{4.14}$$

The error matrix, $\mathbf{E}$, can be decomposed into the sum of at most $N-1$ terms of the form $\lambda_i \mathbf{E}_i$, where λ_i is a scalar and $\mathbf{E}_i$ is a differential matrix (that is one with zero row sums) with the following properties.

$$-1 \leqslant \lambda_i < 1$$
$$\mathbf{E}_i^2 = \mathbf{E}_i$$
$$\mathbf{E}_i \mathbf{E}_j = \mathbf{S} \mathbf{E}_i = 0 \qquad j \neq i$$

There is a λ_i term corresponding to the real part of each distinct eigenvalue of $\mathbf{P}$ except the eigenvalue 1. The full decomposition of $\mathbf{P}$ and $\mathbf{P}^n$ is as follows.

$$\mathbf{P} = \mathbf{S} + \sum_{i=2}^{k \leqslant N} \lambda_i \mathbf{E}_i$$
$$\mathbf{P}^n = \mathbf{S} + \sum_{i=2}^{k \leqslant N} \lambda_i^n \mathbf{E}_i$$

A detailed analysis of Markov processes in terms of the eigenvalue spectrum is given by e.g. Cox and Miller 1965. For present purposes it is sufficient to consider the limiting sum of the error terms, which is represented by the fundamental matrix.

4.2.3 The Fundamental Matrix

The fundamental matrix $\mathbf{Y}$ is defined by

$$\mathbf{Y} = (\mathbf{I} - \mathbf{E})^{-1} - \mathbf{S}$$

$\mathbf{Y}$ always exists. This definition differs from Kemény and Snell whose 'fundamental matrix' is $Y + S$. Alternative expressions for $\mathbf{Y}$ are

$$\mathbf{Y} = -\mathbf{S} + \sum_{n=0}^{\infty} \mathbf{E}^n = (\mathbf{I} - \mathbf{S}) + \mathbf{E} + \mathbf{E}^2 + \mathbf{E}^3 + \ldots \qquad (4.16)$$

$$\mathbf{Y} = \sum_{n=0}^{\infty} (\mathbf{P}^n - \mathbf{S}) \qquad (4.17)$$

From equation 4.17 we see that the fundamental matrix has the following physical interpretation. For a system which starts in state i, the ijth element of the matrix $\mathbf{P}^n - \mathbf{S}$, equation 4.17, gives the probability that the system is in state j at stage n minus the limiting probability that the system is in state j at a randomly chosen stage. The ijth element of the fundamental matrix, therefore, gives the mean number of 'extra visits' to state j by a system which starts in state i and continues indefinitely. Extra visits are visits additional to or falling short of those occurring at the steady state or limiting average rate.

From equation 4.17 we see that $\mathbf{Y}$ has zero row sums and that

$$\mathbf{SY} = \mathbf{YS} = \mathbf{0} \tag{4.18}$$

Also by use of equations 4.8, 4.9, 4.11, 4.12 and 4.14 we derive the following results,

$$(\mathbf{I} - \mathbf{P}) = (\mathbf{I} - \mathbf{S})\,(\mathbf{I} - \mathbf{E}) = (\mathbf{I} - \mathbf{E})\,(\mathbf{I} - \mathbf{S}) \tag{4.19}$$

$$(\mathbf{I} - \mathbf{P})\,(\mathbf{I} - \mathbf{E})^{-1} = (\mathbf{I} - \mathbf{S}) = (\mathbf{I} - \mathbf{E})^{-1}(\mathbf{I} - \mathbf{P}) \tag{4.20}$$

but $(\mathbf{I} - \mathbf{P})\mathbf{S} = \mathbf{0}$ therefore

$$(\mathbf{I} - \mathbf{P})\mathbf{Y} = (\mathbf{I} - \mathbf{S}) = \mathbf{Y}(\mathbf{I} - \mathbf{P}) \tag{4.21}$$

$$\mathbf{PY} = \mathbf{YP} = \mathbf{Y} - \mathbf{I} + \mathbf{S} \tag{4.22}$$

and for any column vector $\mathbf{c}$ such that $(\mathbf{I} - \mathbf{P})\mathbf{c} = \mathbf{0}$ we have $\mathbf{Y}(\mathbf{I} - \mathbf{P})\mathbf{c} = \mathbf{0}$ and hence $(\mathbf{I} - \mathbf{S})\mathbf{c} = \mathbf{0}$. Thus

$$(\mathbf{I} - \mathbf{P})\mathbf{c} = \mathbf{0} \Leftrightarrow (\mathbf{I} - \mathbf{S})\mathbf{c} = \mathbf{0} \tag{4.23}$$

4.3 COMPUTATION OF THE FIXED AND FUNDAMENTAL MATRICES

In the solution of infinite stage Markov decision problems it is not essential to be able to compute the fixed and fundamental matrices associated with any given transition probability matrix. Nevertheless, we now perform some such calculations in order to provide further insight into the nature of these matrices and to provide a basis for subsequent discussion of the concept of bias optimality. In this section we consider only the case where the given transition probability matrix has a single recurrent chain. In this case the rows of the fixed matrix are identical. We have encountered an example of this in the slotting machine (Table 4.1) where both rows of $\mathbf{P}^n$ approach the same limit. For a unichain system the limiting probability of finding the system in state j at a randomly chosen stage is independent of the initial state i. The computational method now described can be extended to multi-chain systems with the aid of a partitioning scheme.

Let $\mathbf{D}$ be a stochastic matrix of the same order as $\mathbf{P}$ but with zero elements everywhere except in the right hand column. ($\mathbf{P}$ is any given unichain transition probability matrix.) $\mathbf{D}$ is the 'normalising' matrix.

$$\mathbf{D} = \begin{bmatrix} 0 & 0 & \dots & 1 \\ 0 & 0 & \dots & 1 \\ \vdots & \vdots & & \vdots \\ 0 & 0 & \dots & 1 \end{bmatrix} \tag{4.24}$$

Now $\mathbf{S}$ has identical rows, $\mathbf{P}, \mathbf{S}$ and $\mathbf{D}$ have unit row sums and $\mathbf{Y}$ has zero row sums, therefore

$$\mathbf{DS} = \mathbf{S}$$
$$\mathbf{PD} = \mathbf{SD} = \mathbf{D}^2 = \mathbf{D} \tag{4.25}$$
$$\mathbf{YD} = \mathbf{0}$$

Define a matrix $\mathbf{Q}$ by

$$\mathbf{Q} = \mathbf{I} + \mathbf{D} - \mathbf{P} \tag{4.26}$$

Then

$$\mathbf{Q}^{-1} = \mathbf{S} + \mathbf{Y} - \mathbf{DY} \tag{4.27}$$

Premultiply equation 4.27 by $\mathbf{D}$ and use equations 4.25 to give

$$\mathbf{S} = \mathbf{DQ}^{-1} \tag{4.28}$$

Premultiply equation 4.27 by $\mathbf{S}$ and subtract the result from equation 4.27 to give

$$\mathbf{Y} = (\mathbf{I} - \mathbf{S})\mathbf{Q}^{-1} \tag{4.29}$$

If we wish to compute $\mathbf{S}$ only it is sufficient to compute the bottom row of $\mathbf{Q}^{-1}$ which is identical to every row of $\mathbf{S}$. To get $\mathbf{Y}$ we require the whole of $\mathbf{Q}^{-1}$.

It is convenient at this point to note that for $\mathbf{P}$ unichain, the fact that $\mathbf{S}$ has identical rows together with statement 4.23 yields the further result,

$$(\mathbf{I} - \mathbf{P})\mathbf{c} = \mathbf{0} \Leftrightarrow \mathbf{c} = c\mathbf{1}, \quad \mathbf{P} \text{ unichain} \tag{4.30}$$

where $\mathbf{1}$ is a column vector of ones.

4.3.1 Examples

Consider the general two state system with transition probability matrix

$$\mathbf{P} = \begin{bmatrix} 1-p & p \\ q & 1-q \end{bmatrix} \tag{4.31}$$

There is a special multichain case $p = q = 0$ for which $\mathbf{P} = \mathbf{I}$, $\mathbf{S} = \mathbf{I}$, $\mathbf{Y} = \mathbf{0}$. Apart from this the system is unichain and using equations 4.26, 4.28 and 4.29 we have

$$\mathbf{Q} = \mathbf{I} + \mathbf{D} - \mathbf{P} = \begin{bmatrix} p & 1-p \\ -q & 1+q \end{bmatrix} \tag{4.32}$$

$$\mathbf{Q}^{-1} = (1/(p+q)) \begin{bmatrix} 1+q & p-1 \\ q & p \end{bmatrix} \tag{4.33}$$

$$\mathbf{S} = \mathbf{DQ}^{-1} = (1/(p+q)) \begin{bmatrix} q & p \\ q & p \end{bmatrix} \tag{4.34}$$

$$\mathbf{Y} = (\mathbf{I} - \mathbf{S})\mathbf{Q}^{-1} = (1/(p+q)^2) \begin{bmatrix} p & -p \\ -q & q \end{bmatrix} \tag{4.35}$$

Consider the case $p = \frac{1}{2}$, $q = 0$, that is,

$$\mathbf{P} = \begin{bmatrix} \frac{1}{2} & \frac{1}{2} \\ 0 & 1 \end{bmatrix}$$

Using equations 4.34 and 4.35 the fixed and fundamental matrices are

$$\mathbf{S} = \begin{bmatrix} 0 & 1 \\ 0 & 1 \end{bmatrix} \qquad \mathbf{Y} = \begin{bmatrix} 2 & -2 \\ 0 & 0 \end{bmatrix} \tag{4.36}$$

Let $\mathbf{S} = [s(i,j)]$, $\mathbf{Y} = [y(i,j)]$. If state j is transient $s(i,j) = 0$; $i = 1$, ..., N. This reflects the fact that the limiting probability of finding the system in a transient state is zero. If state j is absorbing then $s(i,j) = 1$ for all states i which communicate with j and with no other recurrent state. Equations 4.36 illustrate this.

If j is a transient state $y(i,j)$ gives the mean total number of visits the system makes to state j given that it starts in state i. In equations 4.36, $y(1, 1) = 2$ and this means that on average two visits are made to state 1 if the system starts in that state. It is easy to see how this comes about. The probability of a visit at stage n is $\frac{1}{2}^n$ for $n = 0$, 1, 2, ... The sum to infinity of these terms is $1/(1 - \frac{1}{2}) = 2$. Thus there are two visits on average. This also implies a shortfall of two visits to state 2, giving $y(1, 2) = -2$.

4.3.2 Slotting Machine

For the slotting machine example introduced earlier we have

$$\mathbf{P} = \begin{bmatrix} 0 \cdot 4 & 0 \cdot 6 \\ 0 \cdot 2 & 0 \cdot 8 \end{bmatrix} \tag{4.1}$$

Comparing with equation 4.31 we have $p = 0 \cdot 6$, $q = 0 \cdot 2$. Equations 4.34 and 4.35 then yield

$$\mathbf{S} = \begin{bmatrix} \frac{1}{4} & \frac{3}{4} \\ \frac{1}{4} & \frac{3}{4} \end{bmatrix} \qquad \mathbf{Y} = \begin{bmatrix} 15/16 & -15/16 \\ -5/16 & 5/16 \end{bmatrix} \tag{4.37}$$

$\mathbf{S}$ has the value anticipated from Table 4.1, and indicates a steady state availability of 75% for the slotting machine. The fundamental matrix indicates that the system makes $-15/16$ extra visits to state 2 if it starts in state 1, and 5/16 extra visits to state 2 if it starts in state 2, extra visits being the visits additional to those occuring at the steady state rate. Thus the situation where the machine is currently running has an advantage equivalent to $1\frac{1}{4}$ intervals of production over the situation where the machine is currently resetting.

4.3.3 Periodic System

A periodic state has the property that the system returns to it at regular intervals. Earlier we met the transition probability matrix

$$\mathbf{P} = \begin{bmatrix} 0 & 1 \\ 1 & 0 \end{bmatrix} \tag{4.38}$$

The system oscillates regularly between states 1 and 2. Comparing equations 4.38 and 4.31 the latter corresponds to the case $p = q = 1$.

Application of equations 4.34 and 4.35 gives,

$$S = \begin{bmatrix} \frac{1}{2} & \frac{1}{2} \\ \frac{1}{2} & \frac{1}{2} \end{bmatrix} \qquad Y = \begin{bmatrix} \frac{1}{4} & -\frac{1}{4} \\ -\frac{1}{4} & \frac{1}{4} \end{bmatrix} \qquad (4.39)$$

In view of the regular oscillation of the system between its states the probability that it is found in a particular state at a randomly chosen stage is $\frac{1}{2}$. This corresponds to the elements of S. For a system with N periodic states $s(i,j) = 1/N$.

The value taken by the fundamental matrix is interesting. The element in the first row and column of $(P^n - S)$ takes the value $\frac{1}{2}$ for n even and $-\frac{1}{2}$ for n odd. The sum of the first n such values is $\frac{1}{2}$ for n even and 0 for n odd. The stagewise average sum is therefore $\frac{1}{4}$ and this is the value of $y(1, 1)$. Similar remarks apply to the remaining $y(i,j)$ terms. Thus the fixed and fundamental matrices take just the appropriate stagewise average values which are required in the analysis of infinite stage deterministic processes, and many results, and in particular the policy iteration algorithms, apply to both probabilistic and deterministic processes.

4.4 STATIONARY MARKOV PROCESSES WITH RETURNS

Consider a system with states $i = 1, \ldots, N$ which undergoes a Markov process with transition probability matrix P. When the system makes a transition from state i to state j let a transition return $c(i, j)$ be generated. The transition returns are assumed to be stationary, that is independent of the stage number. By comparison with equation 3.7 we define the stage return associated with state i by

$$r(i) = \sum_{j=1}^{N} p(i,j)c(i,j) \qquad (4.40)$$

The stage returns can be listed in the form of a column vector r

$$r = [r(i)].$$

Suppose that the system starts in state i and undergoes a process that continues for n stages. The mean total return generated is denoted by

$f(n, i)$ and we have the following recurrence relation, compare equation 3.8,

$$f(n, i) = r(i) + \sum_{j=1}^{N} p(i, j) f(n-1, j), \qquad i = 1, \ldots, N. \quad (4.41)$$

The $f(n, i)$ terms are the state values at stage n. They can be listed as a column vector $\mathbf{f}(n)$

$$\mathbf{f}(n) = [f(n, i)]$$

In matrix notation the set of equations 4.41 is,

$$\mathbf{f}(n) = \mathbf{r} + \mathbf{P}\mathbf{f}(n-1) \quad (4.42)$$

Equation 4.42 implies

$$\mathbf{f}(n) = \mathbf{r} + \mathbf{P}\mathbf{r} + \mathbf{P}^2 \mathbf{f}(n-2) \quad (4.43)$$

$$\mathbf{f}(n) = (\mathbf{I} + \mathbf{P} + \mathbf{P}^2 + \ldots + \mathbf{P}^{n-1})\mathbf{r} + \mathbf{P}^n \mathbf{f}(0) \quad (4.44)$$

Now consider the limiting value function denoted by $\mathbf{f}_a(n)$ and defined by

$$\mathbf{f}_a(n) = n\mathbf{S}\mathbf{r} + \mathbf{Y}\mathbf{r} + \mathbf{S}\mathbf{f}(0) \quad (4.45)$$

In the completely ergodic case $\mathbf{f}_a(n)$ is the asymptotic form of $\mathbf{f}(n)$. In the periodic case it is the form about which $\mathbf{f}(n)$ oscillates.

Define column vectors g and w, known respectively as the gain and bias values, by the equations

$$\mathbf{g} = \mathbf{S}\mathbf{r}, \quad \mathbf{w} = \mathbf{Y}\mathbf{r} \quad (4.46)$$

The ith element of g, $g(i)$, is the asymptotic increase in the value of state i per stage, hence the name gain. The bias value $w(i)$ of state i is the mean total return generated at the 'extra visits' which the system makes to the various states when it starts in state i. Thus it is the mean total return generated over and above the steady state or limiting component $\mathbf{S}\mathbf{r}$ and the terminal component $\mathbf{S}\mathbf{f}(0)$.

For a unichain system the rows of $\mathbf{S}$ are identical row vectors $\mathbf{s}'$. g therefore has the form $g\mathbf{1}$ where g is a scalar given by

$$g = \mathbf{s}'\mathbf{r} \quad (4.47)$$

Using definitions 4.46, equation 4.45 can be written in the form

$$\mathbf{f}_a(n) = n\mathbf{g} + \mathbf{w} + \mathbf{S}\mathbf{f}(0) \quad (4.45a)$$

4.4.1 Slotting Machine Example

To illustrate a Markov process with returns we first add a return structure to the slotting machine problem. The transition probability matrix is

$$\mathbf{P} = \begin{bmatrix} 0.4 & 0.6 \\ 0.2 & 0.8 \end{bmatrix} \tag{4.1}$$

If a machine is resetting at the start of an interval and then again resetting at the start of the next a cost of 2 units is incurred. If the machine is resetting at the start of an interval and then is running at the start of the next there is a return of 3 units. If the machine is running at the start of an interval and then again running at the start of the next the return is 6 units. Finally, if the machine is running at the start of an interval and then resetting at the start of the next the return is 3 units. This data is summarised by the transition return matrix

$$\mathbf{C} = \begin{bmatrix} -2 & 3 \\ 3 & 6 \end{bmatrix} \tag{4.48}$$

Using equation 4.40, the stage returns are

$$\mathbf{r} = \begin{bmatrix} 1.0 \\ 5.4 \end{bmatrix} \tag{4.49}$$

The fixed and fundamental matrices for the slotting machine are given in equations 4.37. Hence

$$\mathbf{g} = \mathbf{Sr} = \begin{bmatrix} \frac{1}{4} & \frac{3}{4} \\ \frac{1}{4} & \frac{3}{4} \end{bmatrix} \begin{bmatrix} 1.0 \\ 5.4 \end{bmatrix} = \begin{bmatrix} 4.3 \\ 4.3 \end{bmatrix} \tag{4.50}$$

The scalar steady state gain is $g = 4.3$.

$$\mathbf{w} = \mathbf{Yr} = (1/16) \begin{bmatrix} 15 & -15 \\ -5 & 5 \end{bmatrix} \begin{bmatrix} 1.0 \\ 5.4 \end{bmatrix} = (1/8) \begin{bmatrix} -33 \\ 11 \end{bmatrix} \tag{4.51}$$

The limiting value function is

$$f_a(n, 1) = 4.3n - 33/8 \tag{4.52}$$

$$f_a(n, 2) = 4.3n + 11/8. \tag{4.53}$$

The approach of the n stage values $\mathbf{f}(n)$ to the asymptotes is illustrated in Figure 4.2. The lines represent equations 4.52 and 4.53 and the points are the exact n stage values, which were calculated from equation 4.41 using the terminal values $\mathbf{f}(0) = \mathbf{0}$. The bias values are the intercepts of the asymptotes with the line $n = 0$.

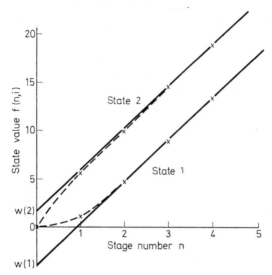

Figure 4.2. Asymptotic behaviour of the state values $f(n, i)$.

4.5 RELATIVE BIAS VALUES

In the preceding section we computed the gain and bias values from the the defining equations $\mathbf{g} = \mathbf{Sr}$, $\mathbf{w} = \mathbf{Yr}$. In the optimisation problems which follow we make use of quantities called relative bias values. The relative bias value vector, $\mathbf{v}$, associated with a transition probability matrix $\mathbf{P}$ and return vector $\mathbf{r}$ is defined by

$$\mathbf{v} = \mathbf{w} + \mathbf{c} = \mathbf{Yr} + \mathbf{c} \qquad (4.54)$$

where $\mathbf{c}$ is any vector such that $(\mathbf{I} - \mathbf{P})\mathbf{c} = \mathbf{0}$.

For a unichain system $\mathbf{c}$ must have the form $c\mathbf{1}$. Thus $\mathbf{v}$ is the vector of bias values determined to within an arbitrary constant in each unichain subset.

The advantage of the relative bias values is that they can be determined more easily than the absolute bias values, and they are often just as useful for determining optimal policies. In fact in the unichain case the absolute bias values are not needed. The relative bias values also provide a method for computing the absolute bias values without determining the fundamental matrix. The equations for determining g, v and w for a unichain Markov process with returns are now developed.

Postmultiply equation 4.21 by the return vector r to give

$$(I-P)Yr = (I-S)r \qquad (4.55)$$

Hence

$$(I-P)w = r-g \qquad (4.56)$$

$$(I-P)v = r-g \qquad (4.57)$$

For a unichain system g has the form $g1$ and hence,

$$(I-P)v + g1 = r \qquad (4.58)$$

Equation 4.58 represents N linear simultaneous equations with $N+1$ unknowns. However, addition of a constant to each relative bias value does not affect the left hand side. We can therefore choose one relative bias value arbitrarily and use equation 4.58 to find the gain and the remaining relative bias values.

If the absolute bias values are required they can be found by computing the fixed matrix S and using the equation

$$w = (I-S)v \qquad (4.59)$$

4.5.1 Slotting Machine Example

To illustrate these results consider the slotting machine example with P and r given by equations 4.1 and 4.49. Equation 4.58 is then as follows

$$\begin{bmatrix} 0\cdot6 & -0\cdot6 \\ -0\cdot2 & 0\cdot2 \end{bmatrix} \begin{bmatrix} v(1) \\ v(2) \end{bmatrix} + \begin{bmatrix} g \\ g \end{bmatrix} = \begin{bmatrix} 1\cdot0 \\ 5\cdot4 \end{bmatrix} \qquad (4.60)$$

Since one of the elements of v is to be chosen arbitrarily we can put $v(2) = 0$. The simultaneous equations then are

$$\left. \begin{array}{r} 0\cdot6v(1) + g = 1\cdot0 \\ -0\cdot2v(1) + g = 5\cdot4 \end{array} \right\} \qquad (4.61)$$

Their solution is

$$\left.\begin{array}{c} g = 4 \cdot 3 \\ v(1) = -5 \cdot 5 \end{array}\right\} \tag{4.62}$$

This confirms the gain found earlier and also indicates that the bias value of state 1 is 5·5 units less than the bias value of state 2. This is in accordance with the absolute bias values obtained from equation 4.51, although of course equation 4.61 gives only relative bias values.

As an illustration of equation 4.59 we use the relative bias values

$$\mathbf{v} = \begin{bmatrix} -5 \cdot 5 \\ 0 \end{bmatrix}$$

and the fixed matrix $\mathbf{S}$ as given in equation 4.37 to determine the absolute bias values, thus

$$\mathbf{w} = (\mathbf{I} - \mathbf{S})\mathbf{v} = \begin{bmatrix} \frac{3}{4} & -\frac{3}{4} \\ -\frac{1}{4} & \frac{1}{4} \end{bmatrix} \begin{bmatrix} -5 \cdot 5 \\ 0 \end{bmatrix} = \begin{bmatrix} -33/8 \\ 11/8 \end{bmatrix} \tag{4.63}$$

This is the same result as obtained in equation 4.51.

4.6 INFINITE STAGE MARKOV DECISION PROBLEMS

Consider a system with states $i = 1, \ldots, N$. In state i there is a set K_i of alternative actions k. If the system is in state i and action k is chosen the system goes to state j with probability $p(i, j, k)$. When the system goes from state i to state j under action k it generates a transition return $c(i, j, k)$. This situation is similar to the finite stage case discussed in Chapter 3, except that we now assume that the transition probabilities are independent of stage number. The row vector of transition probabilities from state i to states $j = 1, \ldots, N$ under action k is denoted by $\mathbf{p}'(i, k)$

$$\mathbf{p}'(i, k) = [p(i, j, k)] \tag{4.64}$$

The stage return (mean transition return) associated with state i and action k is $r(i, k)$ given by

$$r(i, k) = \sum_{j=1}^{N} p(i, j, k)c(i, j, k) \tag{4.65}$$

A *policy* is a set of actions, one for each of states at a given stage. Every such set of actions constitutes a valid policy for the system. In finite stage problems we were concerned with finding optimal plans, a plan being a set of actions, one for each state at each of a number of stages. Thus a plan consists of sequence of policies in which several different policies may occur.

In infinite stage problems we assume that the system is to be operated under the same policy at every stage. Our aim is to find a policy which, if repeated indefinitely, will have better limiting properties than other policies. A plan in which a given policy is repeated at every stage is called a stationary plan.

Some authors have highlighted n stage periodic problems in which a nonstationary plan is superior to a stationary plan for all n. White 1969, page 93, gives an example of this; however, the nonstationary plan can only be implemented if it is always known whether the (infinite) number of remaining stages is odd or even and in practice this would not be known. In the absence of this information the best plan is a stationary one. Mine and Osaki 1970 show that stationary plans are superior to randomised plans. There is thus ample justification for confining our attention to stationary plans.

A general policy is denoted by a column vector $\mathbf{k}$. The ith element of $\mathbf{k}$ is the number of the action specified for state i under that policy. $\mathbf{P}(\mathbf{k}) = [p(i, j, k)]$ is the transition probability matrix for the system under policy $\mathbf{k}$. The column vector of stage returns associated with policy $\mathbf{k}$ is $\mathbf{r}(\mathbf{k}) = [r(i, k)]$. Note that where a quantity depends on a single action k it is written as a function of k, but a quantity which depends on all the actions of a policy $\mathbf{k}$ is written as a function of $\mathbf{k}$. The fixed and fundamental matrices and the gain, bias values and relative bias values for the system under policy $\mathbf{k}$ are denoted respectively by $\mathbf{S}(\mathbf{k})$, $\mathbf{Y}(\mathbf{k})$, $\mathbf{g}(\mathbf{k})$, $\mathbf{w}(\mathbf{k})$, $\mathbf{v}(\mathbf{k})$.

In equation 4.45 we introduced the limiting value function $\mathbf{f}_a(n)$ for a Markov process with returns. The limiting value function under policy $\mathbf{k}$ will be denoted by $\mathbf{f}_a(n, \mathbf{k})$. By comparision with equation 4.45a we have

$$f_a(n, \mathbf{k}) = n\mathbf{g}(\mathbf{k}) + \mathbf{w}(\mathbf{k}) + \mathbf{S}(\mathbf{k})\mathbf{f}(0) \qquad (4.66)$$

4.6.1 Optimisation Criteria

In infinite stage Markov programming the aim is to maximise the limiting values $f_a(n, k)$ over the available policies. The limiting values will of course tend to infinity with n under every policy. However, a policy with a larger gain will increase more rapidly than a policy with a smaller gain. Thus our first priority is to find a policy which has as large a gain as possible.

More specifically, we say that a policy **m** is *gain optimal* if,

$$\mathbf{g(m)} \geqslant \mathbf{g(k)} \quad \text{for all policies } \mathbf{k}$$

$\mathbf{g(m)} \geqslant \mathbf{g(k)}$ means that $g(i, m) \geqslant g(i, k)$ for $i = 1, \ldots, N$. The fact that actions can be chosen independently to form policies ensures that a gain optimal policy will exist even in the multichain case where **g** is not necessarily scalar.

It may well happen that several policies have the same gain in a given problem. For example in a unichain system with transient states the gain will not be affected by the actions chosen in the transient states. In considering the problem of tie gain optimal policies we shall leave the terminal values $\mathbf{f}(0)$ out of account, that is we assume $\mathbf{f}(0) = \mathbf{0}$. The limiting value function for a gain optimal policy **m** is then

$$\mathbf{f}_a(n, \mathbf{m}) = n\mathbf{g(m)} + \mathbf{w(m)} \tag{4.67}$$

In resolving ties between gain optimal policies it is advantageous to choose the policy with the largest bias values. We define a policy **z** as bias optimal if

$$\mathbf{g(z)} = \mathbf{g(m)} \geqslant \mathbf{g(k)} \quad \text{for all policies } \mathbf{k}$$

$$\mathbf{w(z)} \geqslant \mathbf{w(m)} \quad \text{for all gain optimal policies } \mathbf{m}$$

The fact that actions can be chosen independently to form policies ensures that a bias optimal policy exists.

4.6.2 Slotting Machine Example

As an example of an infinite stage Markov decision problem we extend the two state slotting machine example to include alternative actions. A slotting machine is either resetting, state $i = 1$, or running, state

$i = 2$. It is easier to reset a machine with new cutters than with part worn ones. On the other hand, cutters are costly and expenditure on them can be kept down by reusing them until they are worn to a certain tolerance level. The data for state 1 given in equations 4.1 and 4.49, and repeated in Table 4.2, relates to the situation where part worn cutters are reused, which we refer to as action $k = 1$ for state 1. Action $k = 2$ for state 1 is to use new cutters at every reset. The transition probability from state i to state j under action k is denoted by $p(i, j, k)$ and the corresponding transition return by $c(i, j, k)$. For action 2 at state 1 the transition probabilities are $p(1, 1, 2) = 0·3$, $p(1, 2, 2) = 0·7$ and the transition returns are $c(1, 1, 2) = -3·3$, $c(1, 2, 2) = 2·7$. The stage return $r(i, k)$ is calculated using equation 4.40 and is $r(1, 2) = 0·9$.

A slotting machine can cease to run satisfactorily for a number of reasons such as running out of screw blanks, tool wear beyond the tolerance level or faulty feeding. The transition probabilities and returns given in equations 4.1 and 4.49 for state 2 are those associated with a mode of operation known as normal feeding. Machines can be operated in a mode known as special feeding in which extra care is taken to ensure that the feed material is of good quality and in good supply. The effect of special feeding is to improve the chances of continued satisfactory running but to increase costs. For state 2 let normal feeding be action 1 and special feeding be action 2. With special feeding the transition probabilities are $p(2, 1, 2) = 0·1$, $p(2, 2, 2) = 0·9$, and the transition returns are $c(2, 1, 2) = 2·2$, $c(2, 2, 2) = 5·2$. The stage return is therefore $r(2, 2) = 4·9$. These data are summarised in Table 4.2. In subsequent sections we shall find gain and bias optimal policies for the slotting machine problem.

4.7 POLICY OPTIMISATION

There are several possible approaches to policy optimisation in infinite stage Markov programming. The main types of algorithm are policy iteration, value iteration and linear programming, but hybrid algorithms can also be devised. Within the confines of the present book it is necessary to treat the optimisation problem selectively. It will be shown how gain and bias optimal policies for unichain undiscounted systems can be found by a hybrid algorithm called the policy-value

Table 4.2. SLOTTING MACHINE PROBLEM: DATA SUMMARY

State i	Action k	Transition probabilities $p(i, 1, k)$	$p(i, 2, k)$	Transition returns $c(i, 1, k)$	$c(i, 2, k)$	Stage returns $r(i, k)$
1 (Resetting)	1 (Old cutters)	0·4	0·6	−2	3	1·0
	2 (New cutters)	0·3	0·7	−3·3	2·7	0·9
2 (Running)	1 (Normal feed)	0·2	0·8	3	6	5·4
	2 (Special feed)	0·1	0·9	2·2	5·2	4·9

iteration algorithm. This is possibly the most efficient algorithm for small and medium sized problems. The value iteration algorithm will then be used to find gain optimal policies for unichain undiscounted systems and bias optimal policies for discounted systems. For further discussion of policy optimisation the reader is referred to Howard 1960, 1971, Mine and Osaki 1970, Denardo 1970, Veinott 1966.

4.8 POLICY-VALUE ITERATION ALGORITHM

4.8.1 Finding a Gain Optimal Policy

The policy-value iteration algorithm (Hastings 1968, 1969) is an accelerated version of Howard's policy iteration algorithm in which intermediate bias values are introduced as policy improvement proceeds. The version presented here applies to systems which are unichain under all policies. The algorithm consists of a repeated cycle of two parts known respectively as the policy evaluation operation (or value determination operation) and the policy improvement routine. In the policy evaluation operation the gain g and relative bias values $\mathbf{v}$ of the current policy are found. In the policy improvement routine either an improved policy is found or the current policy is shown to be optimal in which case the procedure stops. The algorithm is now described in detail and then used to solve the slotting machine problem.

Policy Evaluation Operation

The algorithm starts with any given policy. In the policy evaluation operation the gain and relative bias values under this policy are determined using the linear simultaneous equations 4.58.

$$(\mathbf{I} - \mathbf{P})\mathbf{v} + g\mathbf{1} = \mathbf{r} \tag{4.58}$$

One of the $v(i)$ terms is set to zero and the equations solved for g and $\mathbf{v}$ which are recorded in the *value table*. Alternatively the algorithm can start with an arbitrary set of relative bias values, e.g. $\mathbf{v} = \mathbf{0}$ on the first cycle.

Policy Improvement Routine

In the policy improvement routine either an improved policy is found or the current policy is shown to be optimal.

Each state is considered in turn. For state 1 we find the action which maximises the test quantity

$$r(1, k) + \mathbf{p}'(1, k)\mathbf{v}$$

where $\mathbf{v}$ is the vector of relative bias values for the current policy, as just calculated in the policy evaluation operation. The maximising action forms part of the new policy. If the action used in the current policy is a tie maximising action then that action is retained.

We next calculate an *intermediate bias value*, denoted $v^I(1)$, which is given by.

$$v^I(1) = \underset{k}{\text{Max}}\ [r(1, k) + \mathbf{p}'(1, k)\mathbf{v}] - g \qquad (4.68)$$

The intermediate bias value is thus the maximised test quantity minus the gain of the current policy. The value table is then updated by replacing the relative bias value of state 1, $v(1)$, by the intermediate bias value $v^I(1)$. This concludes the processing of state 1 at this cycle.

We then go to state 2 where the procedure is similar to that at state 1, but uses the current value table which contains the relative bias values $v(i)$ for $i = 2, \ldots, N$ and the intermediate bias value $v^I(1)$. In general for state i the routine is as follows. The action is found which maximises the test quantity

$$r(i, k) + \sum_{j=1}^{i-1} p(i, j, k)v^I(j) + \sum_{j=i}^{N} p(i, j, k)v(j) \qquad (4.69)$$

This action is noted as part of the new policy, the existing action being retained if it is a tie maximising action. The intermediate bias value of state i is then calculated from the equation

$$v^I(i) = \underset{k}{\text{Max}} \left[r(i, k) + \sum_{j=1}^{i-1} p(i, j, k)v^I(j) + \sum_{I=i}^{N} p(i, j, k)v(j) \right] - g \qquad (4.70)$$

The value table is updated by replacing $v(i)$ by $v^I(i)$.

When every state has been processed in this way the new policy is compared with the existing policy. If the two are the same then the existing policy is optimal and the procedure stops. Otherwise we return to the policy evaluation operation taking the new policy as the current policy. A flow chart for the algorithm is shown in Figure 4.3.

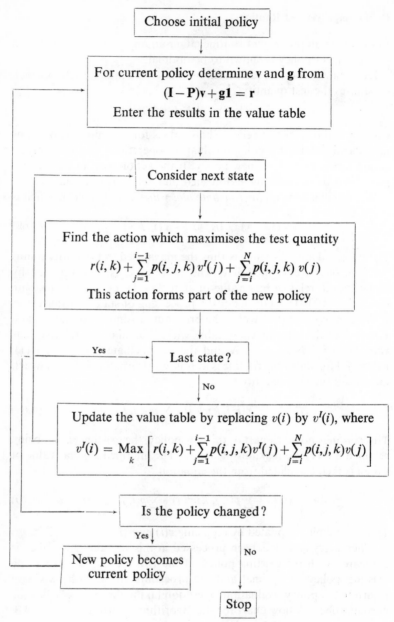

Figure 4.3. Flow chart for the policy-value interation algorithm

4.8.2 Bounds on the Gain

Convergence of the algorithm can be monitored by computing a bound on the optimal gain. Let g be the gain of the current policy and g^* be the optimal gain. Hastings 1971 shows that the optimal gain is bounded by

$$g^* \leqslant g + \underset{i}{\text{Max}} \ [v^I(i) - v(i)] \qquad (4.71)$$

4.8.3 Slotting Machine Example

The data for the slotting machine problem is given in Table 4.2. We now find a gain optimal policy by policy-value iteration.

Let the initial policy be to take action 1 in each state. We enter the first policy evaluation operation with this policy. Using the data of Table 4.2 in equations 4.58 we get

$$\left. \begin{array}{l} (1 - 0 \cdot 4)v(1) - 0 \cdot 6v(2) + g = 1 \cdot 0 \\ -0 \cdot 2v(1) + (1 - 0 \cdot 8)v(2) + g = 5 \cdot 4 \end{array} \right\} \qquad (4.72)$$

One of the $v(i)$ can be chosen arbitrarily so we put $v(2) = 0$. The solution of the equations 4.72 is then

$$\left. \begin{array}{l} v(1) = -5 \cdot 5 \\ v(2) = 0 \\ g = 4 \cdot 3 \end{array} \right\} \qquad (4.73)$$

These results (which were obtained earlier in equations 4.60 to 4.62) form the first value table.

Next we enter the policy improvement routine. At state 1 the values of the test quantity under actions 1 and 2 respectively are

$$\left. \begin{array}{l} r(1, 1) + \sum_{j=1}^{2} p(1, j, 1)v(j) = 1 \cdot 0 + 0 \cdot 4 \times -5 \cdot 5 + 0 \cdot 6 \times 0 = -1 \cdot 2 \\ r(1, 2) + \sum_{j=1}^{2} p(1, j, 2)v(j) = 0 \cdot 9 + 0 \cdot 3 \times -5 \cdot 5 + 0 \cdot 7 \times 0 = -0 \cdot 75 \end{array} \right\} \qquad (4.74)$$

The larger test quantity results from action 2 which is therefore part of the new policy.

The intermediate bias value $v^I(1)$ is found by subtracting g from the maximal test quantity,

$$v^I(1) = -0.75 - 4.3 = -5.05 \qquad (4.75)$$

This value replaces $v(1)$ in the value table which becomes

$$\left.\begin{array}{rcr} v^I(1) = & -5.05 \\ v(2) = & 0 \\ g = & 4.3 \end{array}\right\} \qquad (4.76)$$

We shall compute a bound on g^* at the end of the cycle, so at this point we note the improvement in the bias value which is

$$v^I(1) - v(1) = 0.45 \qquad (4.77)$$

Continuing to state 2, the values of the test quantity under actions 1 and 2 respectively are

$$\begin{aligned} v^I(2, 1) + p(2, 1, 1)v(1) + p(2, 2, 1)v(2) &= 5.4 + 0.2 \times -5.05 + 0.8 \times 0 \\ &= 4.390 \end{aligned}$$

$$\begin{aligned} r(2, 2) + p(2, 1, 2)v(1) + p(2, 2, 2)v(2) &= 4.9 + 0.1 \times -5.05 + 0.9 \times 0 \\ &= 4.395 \end{aligned}$$

The larger value of the test quantity results from action 2 which is therefore part of the new policy. This is the last state so it is not necessary to update the value table, however, we calculate the improvement in the bias value for the purpose of computing bounds. Thus we have

$$v^I(2) - v(2) = 4.395 - 4.3 - 0 = 0.095. \qquad (4.78)$$

The policy improvement routine is now complete. The new policy is action 2 in each state. Before returning to the policy evaluation operation we calculate the upper bound on the optimal gain. This is given by

$$g + \underset{i}{\text{Max}}\,[v^I(i) - v(i)] = 4.3 + \text{Max}\,[0.45 : 0.095] = 4.75 \qquad (4.79)$$

Thus the optimal gain is known to be not greater than 4.75.

Second Cycle

We now re-enter the policy evaluation operation with the new policy, which is action 2 in each state. The simultaneous equations are

$$
\left.\begin{array}{l}
(1-0{\cdot}3)v(1)-0{\cdot}7v(2)+g = 0{\cdot}9 \\
-0{\cdot}1v(1)+(1-0{\cdot}9)v(2)+g = 4{\cdot}9
\end{array}\right\} \tag{4.80}
$$

Again we put $v(2) = 0$. The solution of equations 4.80 is then

$$
\left.\begin{array}{l}
v(1) = -5 \\
v(2) = 0 \\
g = 4{\cdot}4
\end{array}\right\} \tag{4.81}
$$

Continuing with the second policy improvement routine we find that action 2 is chosen at state 1. At state 2 there is a tie between actions 1 and 2. Since action 2 is part of the current policy it is retained. The policy is thus unchanged from the previous cycle and is gain optimal. The optimal gain is 4.4 units.

4.8.4 Finding a Bias Optimal Policy

In the slotting machine example just completed there were tie gain optimal actions in the final policy improvement routine. There are therefore tie gain optimal policies. We now show how to find a bias optimal policy from among the gain optimal policies. The methods used here are essentially those proposed by Veinott 1966 and modified by Denardo 1970.

A bias optimal policy is found by applying policy-value iteration to a subproblem of the original Markov decision problem. In the sub-problem,

1. Only gain optimal actions are considered.
2. The stage returns are replaced by $-\mathbf{v}^*$, where $\mathbf{v}^*$ is the vector of relative bias values under a gain optimal policy.

A policy which is gain optimal for the subproblem is bias optimal for the original problem.

In the slotting machine example the subproblem contains action 2 at state 1 and actions 1 and 2 at state 2. The transition probabilities are as given in Table 4.2. The relative bias values of a gain optimal policy have already been computed in the final gain optimisation cycle and are given in equation 4.81. The stage returns of the subproblem are therefore 5 for state 1, and 0 for state 2 the latter value applying to both actions. The subproblem data is summarised in Table 4.3.

Table 4.3. SLOTTING MACHINE BIAS OPTIMISATION: THE SUBPROBLEM

State i	Action k	Transition Probabilities $p(i, 1, k)$	$p(i, 2, k)$	Modified Returns $-v^*(i)$
1	1	0·4	0·6	5
2	1	0·2	0·8	0
2	2	0·1	0·9	0

We shall denote the gain of the subproblem by h. The bias values of the subproblem will be referred to as bias discriminators since they are used in the selection of a bias optimal policy, and will be denoted by $\mathbf{u} = [u(i)]$.

In the policy evaluation operation of the subproblem we determine the bias discriminators $\mathbf{u}$ and gain h of the current policy by solving the equations

$$(\mathbf{I} - \mathbf{P})\mathbf{u} + h\mathbf{1} = -\mathbf{v}^* \qquad (4.82)$$

one of the bias discriminators being chosen arbitrarily.

In the policy improvement routine the stage returns are now the same for every action at a given state and can be ignored. Let $u^I(i)$ be the intermediate bias discriminator at state i. In general we find the action within the subproblem which maximises

$$\sum_{j=1}^{i-1} p(i, j, k) u^I(j) + \sum_{j=i}^{N} p(i, j, k) u(j) \qquad (4.83)$$

We then update the value table with $u^I(i)$ where

$$u^I(i) = \underset{k \text{ in subproblem}}{\text{Max}} \left[\sum_{j=1}^{i-1} p(i, j, k) u^I(j) + \sum_{j=i}^{N} p(i, j, k) u(j) \right] - h \qquad (4.84)$$

4.8.5 Slotting Machine Example

We take the gain optimal policy $\mathbf{k} = \begin{bmatrix} 2 \\ 2 \end{bmatrix}$ as the starting policy of the bias optimisation phase. The first policy evaluation operation gives

$$\left. \begin{array}{l} u(1) = 6 \cdot 25 \\ u(2) = 0 \\ h = 0 \cdot 625 \end{array} \right\} \tag{4.85}$$

In the policy improvement routine there is only one action at state 1. This is necessarily optimal and processing of this state leaves the value table unchanged. At state 2 the values of the test quantity under actions 1 and 2 respectively are

$$\left. \begin{array}{l} 0 \cdot 2 \times 6 \cdot 25 + 0 \cdot 8 \times 0 = 1 \cdot 25 \\ 0 \cdot 1 \times 6 \cdot 25 + 0 \cdot 9 \times 0 = 0 \cdot 625 \end{array} \right\} \tag{4.86}$$

Action 1 gives the higher trial value and the new policy is $\mathbf{k} = \begin{bmatrix} 2 \\ 1 \end{bmatrix}$

A further cycle confirms that this policy is gain optimal for the subproblem and hence bias optimal for the original problem.

To demonstrate that policy $\mathbf{k}' = [2 \quad 1]$ has greater absolute bias values than policy $[2 \quad 2]$ we calculate the bias values of both policies. This is most easily done by computing the fixed matrices of each policy from equation 4.34 and then computing the bias values from equation 4.59. The results are as follows

Policy [2 2]	Policy [2 1]
$\mathbf{S} = \begin{bmatrix} 1/8 & 7/8 \\ 1/8 & 7/8 \end{bmatrix}$	$\mathbf{S} = \begin{bmatrix} 1/7 & 6/7 \\ 1/7 & 6/7 \end{bmatrix}$
$\mathbf{w} = (\mathbf{I} - \mathbf{S})\mathbf{v} = \begin{bmatrix} -35/8 \\ 5/8 \end{bmatrix}$	$\mathbf{w} = (\mathbf{I} - \mathbf{S})\mathbf{v} = \begin{bmatrix} -30/7 \\ 5/7 \end{bmatrix}$

The limiting total return of policy $[2 \quad 1]$ is $5/56$ units greater than for policy $[2 \quad 2]$. The bias difference between gain optimal policies is always scalar for unichain systems.

The advantage of a bias optimal policy over other gain optimal policies may seem unimportant in regard to an infinite stage process. However, the infinite stage model is often chosen not to represent a truly infinite process but rather one which terminates randomly. The extra rewards are generated early in the process and might as well be reaped as rejected.

4.9 INFINITE STAGE MARKOV PROGRAMMING BY VALUE ITERATION

In Chapter 3 the value iteration algorithm was used to determine optimal plans for finite stage Markov decision problems. Consider a system which when action k is used in state i generates a stage return $r(i, k)$ and moves to state j with probability $p(i, j, k)$. The mean total return generated over n stages from state i under an optimal plan is $f(n, i)$ which obeys the recurrence relation

$$\left. \begin{aligned} f(n, i) &= \text{Max}_k \left[r(i, k) + \sum_{j=1}^{N} p(i, j, k) f(n-1, j) \right] \\ f(0, i) &\quad \text{given}; \quad i = 1, \ldots, N. \end{aligned} \right\} \quad (4.87)$$

The use of equation 4.87 to find gain optimal policies for undiscounted infinite stage, stationary Markov decision problems is now considered. Attention is confined to unichain aperiodic systems. Convergence properties of the algorithm are discussed by Brown 1965 (undiscounted case) and Shapiro 1968 (discounted case). Odoni 1969 establishes bounds on the optimal gain and Hastings 1971 shows that these bounds also apply to the policy found at iteration stage n. These bounds provide a basis for determining when the gain of the current policy is close enough to the optimum for iteration to be discontinued. These bounds are now given and their use illustrated.

4.9.1 Bounds on the Gain

Consider a system which is unichain and aperiodic under every policy and which has optimal gain g^*. Suppose that we carry out value iteration in accordance with equation 4.87. Let $\theta(n, i)$ denote the difference between the value of state i at stages n and $n-1$

$$\theta(n, i) = f(n, i) - f(n-1, i) \tag{4.88}$$

Let the smallest and largest of these value difference terms be denoted by $\theta_L(n)$ and $\theta_U(n)$ respectively

$$\theta_L(n) = \underset{i}{\text{Min}} \, [\theta(n, i)], \quad \theta_U(n) = \underset{i}{\text{Max}} \, [\theta(n, i)] \tag{4.89}$$

The optimal gain g^* and the gain g of the policy found at stage n are bounded by

$$\theta_L(n) \leqslant g \leqslant g^* \leqslant \theta_U(n) \tag{4.90}$$

Hence

$$g^* - g \leqslant \theta_U(n) - \theta_L(n) \tag{4.91}$$

Inequality 4.91 provides a stopping rule for the value iteration algorithm. The right hand side may reach zero for finite n. If it does a gain optimal policy has been found. Normally the procedure is terminated when the right hand side of inequality 4.91 is small enough for practical purposes. A numerical example is given in the next section. Computational experience suggests that convergence is rapid if the relative terminal values (i.e. the values to within an arbitrary constant) are close to the relative bias values under an optimal policy. Although these values are unknown they can often be estimated fairly well.

4.9.2 Slotting Machine Example

Application of the value iteration algorithm to the slotting machine problem for which the data is given in Table 4.2 yields the results shown in Table 4.4. The terminal values used are $f(0, 1) = f(0, 2) = 0$. At stage 5 the policy is $\mathbf{k}'(5) = [2 \quad 1]$. From the value differences listed in Table 4.4 we know that the gain g of this policy and the optimal gain lie between the following bounds

$$4 \cdot 39958 \leqslant g \leqslant g^* \leqslant 4 \cdot 40012$$

From the policy-value iteration results we known that in fact $g = g^* = 4.4$ and that policy $[2 \quad 1]$ is bias optimal. It is fortuitous that the policy thus found is bias rather than merely gain optimal.

10*

Table 4.4. SLOTTING MACHINE PROBLEM: SOLUTION BY VALUE ITERATION

Stage n	State i	Optimal Action k	Optimal Value f(n, i)	Value Difference θ(n, i)
0	1		0	
0	2		0	
1	1	1	1·0	1·0
1	2	1	5·4	5·4
2	1	2	4·98	3·98
2	2	1	9·92	4·52
3	1	2	9·338	4·358
3	2	1	14·332	4·412
4	1	2	13·7338	4·3958
4	2	1	18·7332	4·4012
5	1	2	18·13338	4·39958
5	2	1	23·13332	4·40012

4.10 DISCOUNTED RETURNS

Consider a system which undergoes a Markov process with transition probability matrix $\mathbf{P}$ and generates returns $\mathbf{r}$. Future returns are discounted with b the discount factor $0 \leqslant b < 1$. The present value of the mean total return generated over n stages from state i is $f(n, i)$. $\mathbf{f}(n) = [f(n, i)]$ is the column vector of n stage present values. This obeys the recurrence relation

$$\mathbf{f}(n) = \mathbf{r} + b\mathbf{P}\mathbf{f}(n-1) \qquad (4.92)$$

Hence

$$\mathbf{f}(n) = (\mathbf{I} + b\mathbf{P} + b^2\mathbf{P}^2 + \ldots + b^{n-1}\mathbf{P}^{n-1})\mathbf{r} + b^n\mathbf{P}^n\mathbf{f}(0) \qquad (4.93)$$

As n tends to infinity $\mathbf{f}(n)$ tends to a limit $\mathbf{w}$ where

$$\mathbf{w} = (\mathbf{I} - b\mathbf{P})^{-1}\mathbf{r} \qquad (4.94)$$

$\mathbf{w}$ is the vector of limiting present values for the infinite stage process and is essentially a vector of bias values.

Consider a Markov decision problem with discounted returns. When action k is used in state i the system generates a return $r(i, k)$ and moves to state j with probability $p(i, j, k)$. The row vector of transition probabilities from state i under action k is $\mathbf{p}'(i, k)$,

$$\mathbf{p}'(i, k) = (p(i, j, k)), \qquad j = 1, \ldots, N.$$

The transition probability matrix and return vector associated with policy $\mathbf{k}$ are $\mathbf{P}(\mathbf{k})$ and $\mathbf{r}(\mathbf{k})$ respectively. The vector of limiting present values under policy $\mathbf{k}$ is $\mathbf{w}(\mathbf{k}) = [w(i, \mathbf{k})]$. A policy $\mathbf{z}$ is optimal if

$$\mathbf{w}(\mathbf{z}) \geqslant \mathbf{w}(\mathbf{k}) \quad \text{for all policies } \mathbf{k}.$$

Optimal policies for discounted Markov decision problems can be found by variants of the many algorithms which apply to the undiscounted case. We consider only the value iteration algorithm.

4.10.1 Solution by Value Iteration

For a Markov decision problem with discounted returns $f(n, i)$ is the mean present value of the returns generated using an optimal plan in an n stage process from state i. $f(n, i)$ obeys the recurrence relation

$$f(n, i) = \operatorname*{Max}_{k} [r(i, k) + b\mathbf{p}'(i, k)\mathbf{f}(n-1)] \tag{4.95}$$

At each value iteration stage bounds on the optimal limiting present values and on the limiting present values of the current policy can be computed. Let $\mathbf{a}$ be the policy found at stage n and $\mathbf{z}$ be an optimal policy. These have limiting present values $\mathbf{w}(\mathbf{a})$, $\mathbf{w}(\mathbf{z})$. The value difference quantities $\theta(n, i)$ defined by equations 4.88 and 4.89 apply to the discounted system with $f(n, i)$ now given by equation 4.95. The bounds on the limiting present values are

$$f(n, i) + (b/(1-b))\theta_L(n) \leqslant w(i, \mathbf{a}) \leqslant w(i, \mathbf{z}) \leqslant f(n, i) + b/(1-b)\theta_U(n) \tag{4.96}$$

An action is suboptimal if it is not part of an optimal policy. The following test for suboptimal actions applies at value iteration stage $n+1$. Action k in state i is suboptimal if

$$r(i, k) + b\mathbf{p}'(i, k)\mathbf{f}(n) < f(n, i) + (b/(1-b))(\theta_L(n) - b\theta_U(n)) \tag{4.97}$$

Actions which fail this test can eliminated from consideration at subsequent value iteration stages. The value iteration procedure and the bounds and suboptimality test just described are illustrated in the following example. Proof of the validity of this test is given by Hastings and Mello (1972), which also describes bounds and suboptimality test for the semi-Markov case.

4.10.2 Discounted Replacement Problem

A machine used in a production process can be either in its first year of life (state 1) or in its second year of life (state 2) or have been overhauled (state 3). If the current machine is in its first year, a new machine may be ordered for delivery at the end of the year (action 1), an overhaul of the current machine arranged for the end of the year (action 2) or no arrangements made at all (action 3). The same action set is available in states 2 and 3, except that action 3 cannot be taken in state 2. There are uncertainties in the delivery of new machines, the availability of overhaul facilities and the extent of wear to a machine in any year. It may therefore turn out to be necessary, for example, to cancel an order for a new machine and have an overhaul done instead. Transition probabilities between the states under the various actions have been assessed, as have the stage returns associated with the various states and actions. The returns represent the mean net income from production less expenditure on resources, including the cost of replacement or overhaul where this is applicable. The data, is shown in Table 4.5. There is no clear planning horizon for the

Table 4.5. DISCOUNTED REPLACEMENT PROBLEM: DATA

Discount factor $b = 0.25$

State i	Action k	Transition Probabilities $p(i, j, k)$			Stage Returns $r(i, k)$
		$j = 1$	2	3	
1	1	0·50	0·25	0·25	8·0
	2	0·06	0·75	0·19	2·75
	3	0·25	0·12	0·63	4·25
2	1	0·50	0·00	0·50	16·0
	2	0·06	0·88	0·06	15·0
3	1	0·25	0·25	0·50	7·0
	2	0·12	0·76	0·12	4·0
	3	0·75	0·06	0·19	4·5

facility, but the product could become obsolete at any time. To reflect this uncertainty a discount factor $b = 0.25$ is used. Determine a policy which maximises the limiting present values.

Solution

We choose the arbitrary terminal values $f(0) = 0$. This means that at stage 1 we choose the actions with the highest stage returns, namely action 1 in each state. This is shown in Table 4.6, where, since the optimisation is straightforward, only the optimal actions and values are shown. The value differences $\theta(n, i)$, are also listed.

Table 4.6. DISCOUNTED REPLACEMENT PROBLEM: VALUE ITERATION STAGE 1

Stage *n*	State *i*	Action *k*	Value $f(n, i)$	Value differences $\varphi(n, i)$
1	1	1	8·0	8·0
1	2	1	16·0	$16\cdot0 = \theta_U(1)$
1	3	1	7·0	$7\cdot0 = \theta_L(1)$

From Table 4.9 we see that $\theta_L(1)$ is 7 units, and $\theta_U(1)$ is 16 units. Since $b = 0.25$ we have $b/(1-b) = 1/3$. The quantities $(b/(1-b))\theta_L(n)$ and $(b/(1-b))\theta_U(n)$ which appear in inequality 4.96 are therefore 2·33 and 5·33 respectively. Also the right hand term of inequality 4·97 is

$$(1/3)(2\cdot33 - 0\cdot25 \times 5\cdot33) = 1\cdot0$$

By adding these various quantities to the state values $f(1)$ we get the bounds and the suboptimality test criteria shown in Table 4.7.

The second value iteration stage is shown in Table 4·8. The suboptimality test criteria, derived in Table 4.7 indicate that actions 2 and 3 in states 1 and 3 are suboptimal. These actions can therefore be eliminated.

At state 2 the suboptimality test criterion is 17. Action 2 is optimal, as is indicated by italics. Action 1, however, cannot be eliminated because it gives a trial value of 17·87 which exceeds the suboptimality

Table 4.7. DISCOUNTED REPLACEMENT PROBLEM: BOUNDS AND SUBOPTIMALITY
TEST CRITERIA AT STAGE 1

State i	Value $f(1, i)$	Lower Bound $f(1, i)+$ $(b/(1-b))\theta_L(1)$	Upper Bound $f(1, i)+$ $(b/(1-b))\theta_U(1)$	Suboptimality test criterion $f(1, i)+$ $(b/(1-b))(\theta_L(1)-b\,\theta_U(1))$
1	8·0	10·33	13·33	9·0
2	16·0	18·33	21·33	17·0
3	7·0	9·33	12·33	8·0

test criterion. The the suboptimality test procedure has reduced the total number of possibly optimal actions from 8 to 4.

The next step is the calculation of new bounds and suboptimality test criterion. From Table 4.8, the smallest and largest value differ-

Table 4.8. DISCOUNTED REPLACEMENT PROBLEM: VALUE ITERATION STAGE 2

Stage n	State i	Action k	Trial value $r(i, k)+b\mathbf{p}'(i, k)\mathbf{f}(n-1)$	Value difference $\theta(n, i)$
2	1	*1*	*10·44*	2·44
2	1	2	6·20 (suboptimal)	
2	1	3	6·33 (suboptimal)	
2	2	1	17·87	
2	2	2	*18·75*	$2·75 = \theta_U(2)$
2	3	*1*	*9·37*	$2·37 = \theta_L(2)$
2	3	2	7·49 (suboptimal)	
2	3	3	6·57 (suboptimal)	

ences at stage 2 are 2·37 and 2·75 respectively. The quantities $(b/(1-b))\,\theta_L(2)$ and $(b/(1-b))\,\theta_U(2)$ are therefore 0·79 and 0·91 respectively, and $(b/(1-b))(\theta_L(2)-b\theta_U(2)) = 0·56$.

Hence the new bounds on the nett present values and the suboptimality test criterion for state 2 are as shown in Table 4.9. The suboptimality test criteria for states 1 and 3 are not needed since only one action remains in each state.

Table 4.9. DISCOUNTED REPLACEMENT PROBLEM: BOUNDS AND SUBOPTIMALITY
TEST CRITERION AT STAGE 2

State i	Value $f(2, i)$	Lower Bound $f(2, i)+$ $(b/(1-b))\theta_L(2)$	Upper Bound $f(2, i)+$ $(b/(1-b))\theta_U(2)$	Suboptimality test criterion $f(2, i)+(b/(1-b))$ $(\theta_L(2)-b\theta_U(2))$
1	10·44	11·23	11·35	
2	18·75	19·54	19·66	19·31
3	9·37	10·16	10·28	

The third value iteration stage is shown in Table 4.10. Action 1 in
state 2 fails the suboptimality test. The optimal policy is thus found
in only three stages and is action 1 in state 1, action 2 in state 2 and
action 1 in state 3. Computation of a further set of bounds shows
that, to three significant figures, the optimal limiting present values
are $w(1) = 11\cdot3$, $w(2) = 19\cdot6$, $w(3) = 10\cdot2$.

Table 4.10. DISCOUNTED REPLACEMENT PROBLEM: VALUE ITERATION STAGE 3

Stage n	State i	Action k	Trial Value $r(i, k)+b\mathbf{p}'(i, k)\,\mathbf{f}(n-1)$	Stage Gain $\theta(n, i)$
3	1	1	11·06	0·62 $= \theta_L(3)$
3	2	1	18·48 (suboptimal)	
3	2	2	19·42	0·67 $= \theta_U(3)$
3	3	1	9·99	0·62

EXERCISES

Q4.1. A repairman at a mining site is either free or busy. If he is
currently busy, the probability that he will still be busy in 5 min time is
0·5. If he is currently free the probability that he will still be free in
5 min time is 0·9. Write down his transition probability matrix, P.
Compute P^2, P^4, P^8 and estimate the limiting proportion of time for

which the repairman is busy. Then compute the fixed matrix and determine the exact proportion.

Answer: 1/6

Q4.2. A cable repair truck has a power driven reel which when full carries 400 m of cable. Repairs involve replacing a 100, 200 or 300 m length of old cable, each length occurring with equal probability. Repairs are carried out by taking new cable from the reel unless the length remaining on the reel is too short. In this case the cable on the reel is removed and scrapped, a new 400 m length is put on the reel and the repair then carried out. Determine the mean length of cable scrapped per repair.

Answer: 400/9 m.

Q4.3. A machine which is inspected at the end of every shift has a cutting tool which can be either 'as new' 'part worn' or 'completely worn'. If at the start of a shift the tool is as new there is probability 0·5 that at the end it will still be as new, probability 0·4 that it will be part worn and probability 0·1 that it will be completely worn. If at the start of a shift the tool is part worn there is probability 0·3 that at the end it will still be part worn and probability 0·7 that it will be completely worn. Completely worn tools must be replaced. Completely or partly worn tools have no scrap value. New tools cost £10. The mean value of production in a shift which starts with a new tool is £100 and in a shift which starts with a part worn tool is £90. Should part worn tools be replaced or reused.

Answer: Part worn tools should be replaced, mean profit £95 per shift.

Q4.4.(a) A pump in a process plant may be either 'as new' or 'reconditioned'. The pump is checked at weekly intervals. If the current pump was as new at the last inspection there is probability 0·5 that it will still be as new at the next, probability 0·3 that it will be possible to recondition and refit it and probability 0·2 that it will have to be scrapped. If the current pump is reconditioned there is probability 0·6 that it will be possible to retain it as it is, probability 0·1 that it can be given a further reconditioning and refitted and probability 0·3 that it will have to be scrapped. Determine the mean time until scrapping if the current pump is (i) as new (ii) reconditioned.

(b) New pumps cost £100 and reconditioning costs £10. Scrap pumps are always replaced immediately by new ones. Determine the steady state cost per week and the bias values associated with new and reconditioned pumps. How much should one be willing to pay for a reconditioned pump.

Answer: 4 weeks, $3\frac{1}{3}$ weeks, £27, −£6·66, £6·66, £86·66.

Q4.5. A man employs a secretary who may resign at the end of any month. The probability that a secretary will resign in the current month is a function of the salary paid to her.

When a secretary leaves, the man employs a temporary whose contract is renewable monthly. At the same time he attempts to find another regular secretary. The probability that he finds a regular secretary in the current month is a function of the amount spent on advertising.

Find the policy which minimises the average cost of providing a secretary for the data in the table, where $p(i, j, k)$ is the probability of transition from state i to state j under action k.

State, i	Action, k	Transition Probabilities $p(i, 1, k)$ $p(i, 2, k)$		Mean Monthly Cost $r(i, k)$
1 = Regular	1 = low salary	0·8	0·2	2
Sec.	2 = high salary	0·9	0·1	3
2 = Temp.	1 = cheap advertising	0·5	0·5	4
	2 = expensive advertising	0·8	0·2	8

Answer: Low salary, cheap advertising.

Average cost 2·57 units per month.

Q4.6. A cargo ship operates between three ports and at each port can arrange to load either of two types of cargo. When cargo k is loaded in port i there is probability $p(i, j, k)$ that the ship will go to port j, and if it does so there is a return $c(i, j, k)$. For the data given in the table determine which cargo should be chosen in each port to maximise the mean return per journey.

Port	Cargo	Transition Probabilities			Transition Returns		
i	k	$p(i,1,k)$	$p(i,2,k)$	$p(i,3,k)$	$c(i,1,k)$	$c(i,2,k)$	$c(i,3,k)$
1	1	0	0·5	0·5	0	7	9
	2	0	0·6	0·4	0	4	3
2	1	0·8	0	0·2	2	0	3
	2	0·4	0	0·6	3	0	2
3	1	0·9	0·1	0	4	20	0
	2	1	0	0	2	0	0

Answer: Load cargo 1 in each port. Mean return 5·84 units per journey.

CHAPTER 5

Semi-Markov Programming

5.1 INTRODUCTION

In Chapters 3 and 4 we considered systems undergoing Markov processes in which transitions between states occurred at discrete time intervals. We now consider processes where the times between transitions are random variables. These are called semi-Markov processes.

Consider a system with a finite number of discrete states $i = 1, \ldots, N$. The system makes transitions between its states. An instant at which the system enters a general state i is known as a renewal point. Let t denote time measured from the last renewal point, that is from the moment of entry to the current state. The time to the next transition is a random variable X_i with distribution function $F_i(t)$,

$$F_i(t) = \text{Prob}\,(X_i \leqslant t)$$

This distribution function depends only on the current state i and the future behaviour of the system depends only on the state it entered at the last renewal point. We shall be concerned with problems where decisions, stage returns and transition probabilities are determined at renewal points.

The probability density of transition from state i to state j at time t is denoted by $\phi_{ij}(t)$. A system may make a transition from state i back into state i and the instant at which this occurs is a renewal point. At the moment of entry into state i the probability that the system will next go to state j is $p(i, j)$ given by

$$p(i, j) = \int_0^\infty \phi_{ij}(t) \, dt \tag{5.1}$$

The sum of the transition probabilities $p(i, j)$ over all states j is unity. The survival function $S_i(t)$ gives the probability that the system remains in state i for longer than time t without making a transition.

$$S_i(t) = 1 - F_i(t) = \text{Prob} \, (X_i > t)$$

The mean duration of visits to state i is denoted $t(i)$ and is often conveniently derived from the equation

$$t(i) = \int_0^\infty S_i(t) \, dt \tag{5.2}$$

In order to evaluate $t(i)$ we frequently have to determine $S_i(t)$ from the probability densities $\phi_{ij}(t)$. We therefore define the cumulative function $H_{ij}(t)$ by

$$H_{ij}(t) = \int_0^t \phi_{ij}(u) \, du \tag{5.3}$$

and note that

$$S_i(t) = 1 - \sum_{j=1}^N H_{ij}(t) \tag{5.4}$$

It is sometimes convenient to use the concept of the probability density of transition from state i to state j *given that the system is headed for state j*. This is denoted by $f_{ij}(t)$ and is related to $\phi_{ij}(t)$ by $f_{ij}(t) = \phi_{ij}(t)/p(i, j)$.
We then have

$$\int_0^\infty f_{ij}(t) \, dt = 1$$

5.1.1 Fuel Pump Example

The following is an example of a semi-Markov process. An aero engine fuel pump can be either new (state $i = 1$) or overhauled (state $i = 2$). New pumps are subject to random failures at a rate of

0·1 per thousand hours and overhauled pumps are subject to random failures at 0·2 per thousand hours. When a failure occurs a new pump is installed, but if a pump survives for 2000 hours it is overhauled. These rules apply to both new and overhauled pumps.

For this process we shall calculate the transition probabilities and the mean duration of visits to each state. Let t denote the operating time in thousands of hours measured from entry to the current state. Consider state 1, new pump. The probability density of transition back into state 1 is given by

$$\left.\begin{aligned}\phi_{11}(t) &= 0\cdot1 \exp(-0\cdot1t) & t \leqslant 2 \\ \phi_{11}(t) &= 0 & t > 2\end{aligned}\right\} \tag{5.5}$$

Because of the cut off at $t = 2$ this is not simply an exponential model. $p(1, 1)$ is given by

$$p(1, 1) = \int_0^\infty \phi_{11}(t)\, dt = \int_0^2 0\cdot1 \exp(-0\cdot1t)\, dt = 0\cdot181 \tag{5.6}$$

Hence

$$p(1, 2) = 1 - p(1, 1) = 0\cdot819$$

A similar analysis for state 2 gives $p(2, 1) = 0\cdot330$, $p(2, 2) = 0\cdot670$.
The survival function for state 1 is

$$\left.\begin{aligned}S_1(t) &= 1 - \int_0^t \phi_{11}(u)\, du = \exp(-0\cdot1t) & t \leqslant 2 \\ S_1(t) &= 0 & t > 2\end{aligned}\right\} \tag{5.7}$$

The mean duration of visits to state 1 is therefore given by

$$t(1) = \int_0^\infty S_1(t)\, dt = \int_0^2 \exp(-0\cdot1t)\, dt = 1\cdot81 \tag{5.8}$$

A similar analysis for state 2 gives $t(2) = 1\cdot65$. The units of $t(i)$ are thousands of hours. In summary the transition probabilities are

$$\mathbf{P} = [p(i, j)] = \begin{bmatrix} 0\cdot181 & 0\cdot819 \\ 0\cdot330 & 0\cdot670 \end{bmatrix} \tag{5.9}$$

The mean durations of visits can be written as a column vector, $\mathbf{t}$,

$$\mathbf{t} = [t(i)] = \begin{bmatrix} 1\cdot81 \\ 1\cdot65 \end{bmatrix} \tag{5.10}$$

5.1.2 Returns

Returns can be associated with each state in a way similar to the discrete case. The return structure is very flexible in semi-Markov programming and can depend on the duration of stay in a state. Let the return density $r_i(t)$ be associated with state i at time t. The mean return associated with state i is given by

$$r(i) = \int_0^\infty r_i(t)\, dt \tag{5.11}$$

5.1.3 Fuel Pump Example

New pumps cost 4 units and overhauling costs 1 unit. A failure costs 10 units including the cost of the new pump required.

For state 1 the mean return $r(1)$ is given by the cost of a failure multiplied by the probability of failure plus the cost of overhaul multiplied by the probability of survival to 2000 hours,

$$r(1) = 10F_1(2) + S_1(2) = 10 \times 0 \cdot 181 + 0 \cdot 819 = 2 \cdot 629 \tag{5.12}$$

For state 2 a similar analysis gives $r = 4 \cdot 0$. Hence the vector of mean returns r is

$$\mathbf{r} = [r(i)] = \begin{bmatrix} 2 \cdot 629 \\ 4 \cdot 0 \end{bmatrix} \tag{5.13}$$

5.2 SEMI-MARKOV PROCESSES AS TWO RETURN SYSTEMS

So far we have discussed semi-Markov processes with emphasis on their continuous time aspect. We now take a different viewpoint. Consider a system which undergoes a Markov process with transition probability matrix **P**. The process is discrete in relation to the number of transitions. When state i is visited two returns are generated. These may be random variables. The first or primary return corresponds to what we have so far regarded as the one and only return and is a cost, profit or yield, etc. The mean primary return is denoted $r(i)$. The secondary return corresponds to what we have so far regarded as the duration of the visit to state i, but it can be interpreted as the consumption of any resource of which time is just one example. The mean secondary return is denoted $t(i)$. The column vectors of mean primary and secondary returns are respectively, $\mathbf{r} = [r(i)]$, $\mathbf{t} = [t(i)]$, $i = 1, \ldots, N$.

5.2.1 The Gain per Unit Time

For a system with transition probability matrix $\mathbf{P}$ and mean primary returns $\mathbf{r}$ the vector of limiting mean primary returns generated per stage is $\mathbf{Sr}$ where $\mathbf{S}$ is the fixed matrix corresponding to $\mathbf{P}$. For a unichain system each row of $\mathbf{S}$ is an identical row vector $\mathbf{s}'$ which we refer to as the fixed vector. The limiting mean primary return per stage, denoted by g_r is given by

$$g_r = \mathbf{s}'\mathbf{r} \tag{5.14}$$

Let the system also generate mean secondary returns $\mathbf{t}$. The limiting mean secondary return per stage, denoted g_t is given by

$$g_t = \mathbf{s}'\mathbf{t} \tag{5.15}$$

If the secondary returns are the durations of visits to the states then g_t is the limiting mean duration of the stages. The ratio g_r/g_t is, therefore, the mean primary return generated per unit time. This quantity will be denoted by g

$$g = g_r/g_t \tag{5.16}$$

g is the gain rate or gain per unit time of the system. In semi-Markov programming interest centres on finding a policy with the maximal gain rate.

5.2.2 Determination of the Gain Rate and Bias Quantities

The gain rate for a given system can be calculated by finding the fixed vector and then using equations 5.14–5.16. However, it is better to use an equation from which both the gain rate and certain bias quantities can be determined. This equation, which will shortly be derived, is a generalised version of equation 4.58 and plays a similar role in policy optimisation.

Consider a system with transition probability matrix **P**, fixed matrix **S**, fundamental matrix **Y**, and primary and secondary returns **r** and **t**. Rearrange equation 4.21 to give

$$\mathbf{S} = \mathbf{I} - (\mathbf{I} - \mathbf{P})\mathbf{Y}$$

Postmultiply by the return vectors to give, for a unichain system

$$\mathbf{Sr} = g_r \mathbf{1} = \mathbf{r} - (\mathbf{I} - \mathbf{P})\mathbf{Yr}$$
$$\mathbf{St} = g_t \mathbf{1} = \mathbf{t} - (\mathbf{I} - \mathbf{P})\mathbf{Yt} \qquad (5.17)$$

Hence on rearrangement of 5.17

$$(g_r/g_t)\,(\mathbf{t} - (\mathbf{I} - \mathbf{P})\mathbf{Yt}) = \mathbf{r} - (\mathbf{I} - \mathbf{P})\mathbf{Yr}$$
$$g\mathbf{t} = \mathbf{r} - (\mathbf{I} - \mathbf{P})\mathbf{Y}(\mathbf{r} - g\mathbf{t}) \qquad (5.18)$$

The vector $\mathbf{Y}(\mathbf{r} - g\mathbf{t})$ plays a role similar to that of the relative bias values in the discrete time model. It is in fact the bias vector of a system whose returns are $\mathbf{r} - g\mathbf{t}$. $r(i) - gt(i)$ is the primary return $r(i)$ less a cost penalty incurred at rate g for time $t(i)$. We denote the vector of these biases by **v**, so that

$$\mathbf{v} = \mathbf{Y}(\mathbf{r} - g\mathbf{t}) = [v(i)], \qquad i = 1, \ldots, N \qquad (5.19)$$

On rearrangement equation 5.18 becomes

$$(\mathbf{I} - \mathbf{P})\mathbf{v} + g\mathbf{t} = \mathbf{r} \qquad (5.20)$$

Compare equations 5.20 and 4.58. 4.58 is a special case of 5.20 in which $\mathbf{t} = \mathbf{1}$. This corresponds to the fact that in the discrete time case the duration of visits to each state is unity.

5.2.3 Fuel Pump Example

For the fuel pump example the transition probabilities, mean durations of visits and mean returns (primary) are shown in equations 5.9, 5.10 and 5.13. We shall use equation 5.20 to determine the gain rate, g, and the bias quantities **v**, the latter being determined to within an arbitrary constant. Putting $v(2) = 0$ equations 5.20 become

$$\left.\begin{array}{r}(1-0{\cdot}181)v(1)+1{\cdot}81g = 2{\cdot}629 \\ -0{\cdot}330v(1)+1{\cdot}65g = 4{\cdot}0\end{array}\right\} \tag{5.21}$$

Solving these equations we get

$$\left.\begin{array}{l}v(1) = -1{\cdot}455 \\ v(2) = 0 \\ g = 2{\cdot}125\end{array}\right\} \tag{5.22}$$

Thus the steady state cost per 1000 hours for the fuel pumps is $2{\cdot}125$ units.

5.3 SEMI-MARKOV DECISION PROBLEMS

In a semi-Markov decision problem there is a set K_i of actions k available in states $i = 1, \ldots, N$. Associated with action k in state i are a mean primary return $r(i, k)$, a mean secondary return $t(i, k)$ and probabilities $p(i, j, k)$ of transition to states $j = 1, \ldots, N$. The policy space A is the cartesian product of all K_i. We consider the problem of finding a policy **m** in A with optimal (maximal or minimal) gain rate.

Only systems which are unichain under all policies are considered here. As in the discrete time case several optimisation techniques are applicable. In fact the same algorithms apply with appropriate modifications. We shall show how gain rate optimal policies can be found by policy-value iteration. Transformation of semi-Markov into discrete time problems is then described. This transformation makes it possible to apply value iteration. Other algorithms are discussed by Denardo and Fox 1968, Howard 1971, Jewell 1963, Mine and Osaki 1970.

5.4 POLICY VALUE ITERATION ALGORITHM

The algorithm is very similar to the discrete time version described in Chapter 4.

Policy Evaluation Operation

The algorithm starts with any given policy. In the policy evaluation operation the gain rate, g, and bias quantities $\mathbf{v}$ for this policy are determined by solving the linear simultaneous equations

$$(\mathbf{I}-\mathbf{P})\mathbf{v}+g\mathbf{t} = \mathbf{r}$$

One of the $v(i)$ is chosen arbitrarily. The resulting gain rate and bias quantities are recorded in the value table.

Policy Improvement Routine

Each state is considered in turn. At state i a new action and intermediate bias quantity, $v^I(i)$, are found. The new action for state i is one which maximises the test quantity

$$r(i, k) - gt(i, k) + \sum_{j=1}^{i-1} p(i, j, k)v^I(j) + \sum_{j=i}^{N} p(i, j, k)v(j)$$

This test quantity differs from the discrete case in that a term involving the secondary return $t(i, k)$ now appears.

The intermediate bias quantity is equal to the maximised test quantity

$$v^I(i) = \underset{k}{\text{Max}} \left[r(i, k) - gt(i, k) + \sum_{j=1}^{i-1} p(i, j, k)v^I(j) + \sum_{j=i}^{N} p(i, j, k)v(j) \right]$$

The value table is updated by replacing the bias quantity $v(i)$ by the intermediate bias quantity $v^I(i)$. This concludes the processing of state i for this cycle.

When every state has been processed in this way the new policy is compared with the existing policy. If the two are the same the existing policy is optimal and the procedure stops. Note that as usual existing actions are retained where possible in the event of ties. If the policies are not the same we return to the policy evaluation operation with the new policy as the current policy. A flow chart for the algorithm is shown in Figure 5.1.

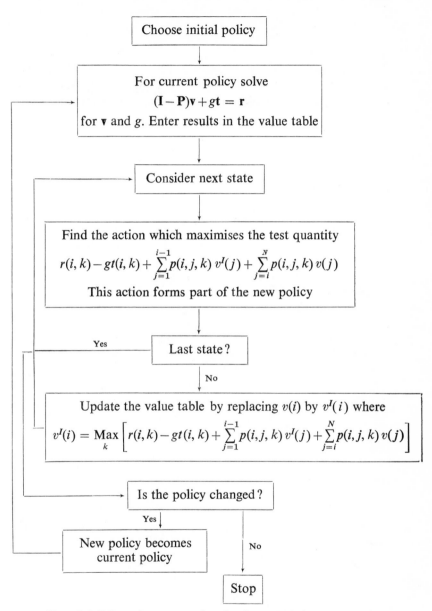

Figure 5.1. Policy value iteration algorithm for semi-Markov programming

5.4.1 Bounds on the Gain Rate

As in the discrete case convergence can be monitored by computing a bound on the optimal gain rate g^*. Let $v(i, k)$ be the value of the test quantity at state i under action k. Hastings 1971 shows that

$$g^* \leqslant g + \underset{i \text{ and } k}{\text{Max}} \left[(v(i, k) - v(i))/t(i, k) \right] \tag{5.23}$$

where g and $v(i)$ are the gain rate and bias quantity for state i under the current policy and $t(i, k)$ is the mean secondary return for state i under action k.

5.4.2 Fuel Pump Problem

We shall extend the fuel pump example into a semi-Markov decision problem which we shall solve by policy value iteration. The situation already described corresponds to a particular operating policy denoted by actions $k = 1$ in each state. The following alternative actions denoted $k = 2$ are available. In state 1 overhaul can be postponed to 2500 hours but 10% of the pumps that reach that age will not be suitable for overhaul and will have to be replaced by new pumps. In state 2 we can similarly postpone overhaul to 2500 hours and in this case 20% of the pumps that reach that age will not be suitable for overhaul.

The transition probabilities, stage returns and mean durations of visits for actions 2 can be computed by extension of the methods given for actions 1. The resulting data are summarised in Table 5.1. We wish to determine the policy which minimises the steady state cost per unit time.

Table 5.1. FUEL PUMP PROBLEM: DATA SUMMARY

State	Action	Transition probabilities		Stage returns	Mean duration of visits
i	k	$p(i, 1, k)$	$p(i, 2, k)$	$r(i, k)$	$t(i, k)$
1	1	0·181	0·819	2·63	1·81
1	2	0·299	0·701	3·22	2·21
2	1	0·330	0·670	4·00	1·65
2	2	0·515	0·485	4·91	1·97

5.4.3 Solution of the Fuel Pump Problem by Policy Value Iteration

Let the initial policy be action $k = 1$ in each state. In the first policy evaluation operation the simultaneous equations are

$$0\cdot819v(1)+1\cdot81g = 2\cdot63$$
$$-0\cdot33v(1)+1\cdot65g = 4\cdot00$$

We have chosen $v(2) = 0$. These equations are in fact the same as 5.21 and their solution is

$$v(1) = -1\cdot455$$
$$v(2) = 0$$
$$g = 2\cdot125$$

Next we enter the policy improvement routine. For state 1 under actions 1 and 2 respectively the values of the test quantity are

$$r(1, 1)-gt(1, 1)+p(1, 1, 1)v(1)+p(1, 2, 1)v(2)$$
$$= 2\cdot63-(2\cdot125\times1\cdot81)+(0\cdot181\times-1\cdot455)+(0\cdot819\times0) = -1\cdot455$$

$$r(1, 2)-gt(1, 2)+p(1, 1, 2)v(1)+p(1, 2, 2)v(2)$$
$$= 3\cdot22-(2\cdot125\times2\cdot21)+(0\cdot299\times-1\cdot455)+(0\cdot701\times0) = -1\cdot912$$

Action 2 minimises the test quantity and is part of the improved policy. The intermediate bias quantity for state 1 is given by the minimal value of the test quantity and is $v^I(1) = -1\cdot912$. This replaces the existing bias quantity $v(1)$ in the value table which becomes

$$v^I(1) = -1\cdot912$$
$$v(2) = 0$$
$$g = 2\cdot125$$

Continuing to state 2, the values of the test quantity under actions 1 and 2 respectively are

$$r(2, 1)-gt(2, 1)+p(2, 1, 1)v^I(1)+p(2, 2, 1)v(2)$$
$$= 4\cdot00-(2\cdot125\times1\cdot65)+(0\cdot330\times-1\cdot912)+(0\cdot67\times0) = -0\cdot136$$

$$r(2, 2)-gt(2, 2)+p(2, 1, 2)v^I(1)+p(2, 2, 2)v(2)$$
$$= 4\cdot91-(2\cdot125\times1\cdot97)+(0\cdot515\times-1\cdot912)+(0\cdot485\times0) = -0\cdot259$$

Action 2 minimises the test quantity and is therefore part of the improved policy. As this is the last state the policy improvement routine is now finished.

At this point a bound on the optimal cost rate can be calculated. In minimisation problems inequality 5.23 becomes

$$g^* \geq g + \underset{i \text{ and } k}{\text{Min}} \left[(v(i, k) - v(i))/t(i, k) \right] \tag{5.24}$$

where as before $v(i, k)$ is the value of the test quantity at state i under action k. Reference to the policy improvement calculations just performed gives the results shown in Table 5.2.

Table 5.2.

i	k	$(v(i, k) - v(i))/t(i, k)$
1	1	$0/1 \cdot 81 = 0$
1	2	$-0 \cdot 457/2 \cdot 21 = -0 \cdot 207$
2	1	$-0 \cdot 136/1 \cdot 65 = -0 \cdot 083$
2	2	$-0 \cdot 259/1 \cdot 97 = -0 \cdot 156$

Action 2 in state 1 gives the minimal entry in the right hand column. From inequality 5.24 we get

$$g^* \geq 2 \cdot 125 - 0 \cdot 207 = 1 \cdot 918$$

The minimal cost rate is not less than $1 \cdot 918$ units per thousand hours. The next policy evaluation operation is entered with the improved policy which is action $k = 2$ in each state. Putting $v(2) = 0$, the simultaneous equations are

$$0 \cdot 701v(1) + 2 \cdot 21g = 3 \cdot 22$$
$$0 \cdot 485v(1) + 1 \cdot 97g = 4 \cdot 91$$

Solving these equations we get the value table

$$v(1) = -1 \cdot 674$$
$$v(2) = 0$$
$$g = 2 \cdot 055$$

The second cycle continues with the policy improvement routine in which it is found that the current policy is optimal and the procedure ends. The solution is to use action 2 in each state.

5.5 DISCOUNTED RETURNS

When returns are continuously discounted at interest rate α the present value of a sum r received after time t is $r \exp(-\alpha t)$. Consider a system which undergoes a semi-Markov process with continuously discounted returns. Suppose that the system enters state i at time $t = 0$. By analogy to equation 5.11, the mean present value of the returns generated at that visit to state i is $r_\alpha(i)$, given by

$$r_\alpha(i) = \int_0^\infty r_i(t) \exp(-\alpha t) \, dt \tag{5.25}$$

Let $w(i)$ denote the mean present value of the returns generated in an infinite duration process which starts in state i at time $t = 0$. $w(i)$ is given by

$$w(i) = \int_0^\infty r_i(t) \exp(-\alpha t) \, dt + \sum_{j=1}^N \int_0^\infty \phi_{ij}(t) w(j) \exp(-\alpha t) \, dt \tag{5.26}$$

Let $p_\alpha(i, j)$ be defined by

$$p_\alpha(i, j) = \int_0^\infty \phi_{ij}(t) \exp(-\alpha t) \, dt \tag{5.27}$$

Using equations 5.25 and 5.27 in 5.26 we get

$$w(i) = r_\alpha(i) + \sum_{j=1}^N p_\alpha(i, j) w(j) \qquad i = 1, \ldots, N \tag{5.28}$$

Note that by definition the Laplace Transform of a function $f(t)$ is

$$\int_0^\infty f(t) \exp(-\alpha t) \, dt$$

Thus $r_\alpha(i)$ and $p_\alpha(i, j)$ are respectively the Laplace Transforms of the functions $r_i(t)$ and $\phi_{ij}(t)$.

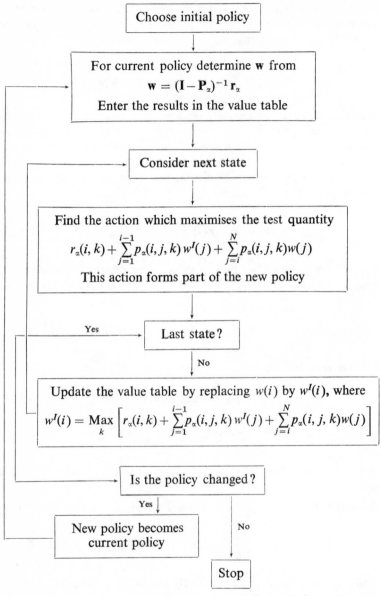

Figure 5.2. Policy value iteration algorithm for discounted semi-Markov programming

Define vectors and matrices $\mathbf{w}$, $\mathbf{r}_\alpha$ and $\mathbf{P}_\alpha$ as follows

$$\left.\begin{aligned}
\mathbf{w} &= [w(i)] \\
\mathbf{r}_\alpha &= [r_\alpha(i)] \\
\mathbf{P}_\alpha &= [p_\alpha(i,j)]
\end{aligned}\right\} \quad i = 1, \ldots, N; \quad j = 1, \ldots, N \quad (5.29)$$

Equations 5.28 in matrix form are

$$\left.\begin{aligned}
\mathbf{w} &= \mathbf{r}_\alpha + \mathbf{P}_\alpha \mathbf{w} \\
\mathbf{w} &= (\mathbf{I} - \mathbf{P}_\alpha)^{-1} \mathbf{r}_\alpha
\end{aligned}\right\} \quad (5.30)$$

Note the close parallel between equation 5.30 and the discrete time equation 4.94.

5.5.1 Semi-Markov Decision Problems with Discounting

In a semi-Markov decision problem with continuous discounting the discounted returns and transition probabilities under action k in state i are $r_\alpha(i, k)$, $p_\alpha(i, j, k)$ for $j = 1, \ldots, N$. The corresponding matrices for policy $\mathbf{k}$ are $\mathbf{r}_\alpha(\mathbf{k})$, $\mathbf{P}_\alpha(\mathbf{k})$. An optimal policy $\mathbf{z}$ maximises the limiting present values, that is, $\mathbf{w}(\mathbf{z}) \geqslant \mathbf{w}(\mathbf{k})$ for every policy $\mathbf{k}$.

An optimal policy can be found by policy-value iteration. The algorithm is a variation of that for the undiscounted case and is outlined in Figure 5.2.

5.5.2 Example

The aeroengine fuel pumps described earlier run for approximately 1000 hours per year. It is decided to allow for discounting by introducing an interest rate of 25% per 1000 hours of service. Determine an optimal policy for the discounted problem.

The first step is to compute the discounted returns and transition probabilities $\mathbf{r}_\alpha$ and $\mathbf{P}_\alpha$ for all policies. That is, we require the discounted equivalent of Table 5.1. Consider the stage return for state 1 and action 1. This was derived without discounting in equation 5.12. The cost of a failure is 10 units and the failure probability density function is $0.1 \exp(-0.1t)$ for $0 \leqslant t \leqslant 2$. The mean present value of the failure costs is therefore

$$\int_0^2 10\times0.1 \exp(-0.1t) \exp(-0.25t) \, dt = 1.44$$

Overhaul costs 1 unit and occurs at time $t = 2$ with probability $S_1(2) = 0.819$. The present value of the mean expenditure on overhaul is therefore

$$0.819 \exp(-0.25\times2) = 0.50$$

The total mean present value of the costs associated with state 1 and action 1 is the sum of the repair and overhaul terms and is

$$r_\alpha(1, 1) = 1.44+0.50 = 1.94$$

The discounted returns for other states and actions are calculated by similar methods and are shown in the right hand column of Table 5.2.

The Laplace transforms of the transition probability density functions for state 1 and action 1 are computed as follows. For transitions from state 1 back into state 1 we have

$$p_\alpha(1, 1, 1) = \int_0^\infty \phi_{11}(t) \exp(-\alpha t) \, dt$$

$$= \int_0^2 0.1 \exp(-0.1t) \exp(-0.25t) \, dt = 0.14$$

The transformed transition probabilities will not sum to unity so $p_\alpha(1, 2, 1)$ must be derived separately. The probability of survival to 2000 hours is $\exp(-0.2) = 0.819$. All pumps which survive to this time enter state 2. This corresponds to an 'impulse function' of magnitude 0.819 at time $t = 2$, and its Laplace Transform is

$$p_\alpha(1, 2, 1) = 0.819 \exp(-\alpha t) = 0.819 \exp(-0.5) = 0.50$$

The remaining data is computed by similar methods and is shown in Table 5.3.

Table 5.3. DISCOUNTED FUEL PUMP PROBLEM: DATA SUMMARY

State	Action	Transformed transition probabilities		Discounted returns
i	k	$p_\alpha(i, 1, k)$	$p_\alpha(i, 2, k)$	$r_\alpha(i, k)$
1	1	0·14	0·50	1·94
1	2	0·21	0·37	2·21
2	1	0·26	0·40	3·04
2	2	0·37	0·23	3·64

The solution of the discounted fuel pump problem by policy value iteration is now outlined.

Policy Evaluation Operation

Let $\mathbf{k} = \begin{bmatrix} 1 \\ 1 \end{bmatrix}$ be the initial policy. The limiting present values for this policy are determined by inserting its transformed transition probabilities and discounted returns in the equation

$$\mathbf{w} = (\mathbf{I} - \mathbf{P}_\alpha)^{-1} \mathbf{r}_\alpha$$

Thus

$$\mathbf{w} = \begin{bmatrix} 0·86 & -0·50 \\ -0·26 & 0·60 \end{bmatrix}^{-1} \begin{bmatrix} 1·94 \\ 3·04 \end{bmatrix} = 2·64 \begin{bmatrix} 0·60 & 0·50 \\ 0·26 & 0·86 \end{bmatrix} \begin{bmatrix} 1·94 \\ 3·04 \end{bmatrix} = \begin{bmatrix} 7·01 \\ 8·22 \end{bmatrix}$$

Policy Improvement Routine

The policy improvement routine is shown in Table 5.4. The improved policy is $\begin{bmatrix} 2 \\ 2 \end{bmatrix}$.

In the second policy evaluation operation we find $\mathbf{w} = \begin{bmatrix} 6·48 \\ 7·92 \end{bmatrix}$. The second policy improvement routine shows this policy to be optimal. This is the same result as was obtained in the undiscounted case. The

Table 5.4. DISCOUNTED FUEL PUMP PROBLEM: POLICY IMPROVEMENT ROUTINE

State i	Action k	Test Quantity $r(i\,k) + \sum_{j=1}^{i-1} p(i, j, k)w^I(j) + \sum_{j=i}^{N} p(i, j, k)w(j)$
1	1	$1{\cdot}94 + 0{\cdot}14 \times 7{\cdot}01 + 0{\cdot}50 \times 8{\cdot}22 = 7{\cdot}01$
1	2	$2{\cdot}21 + 0{\cdot}21 \times 7{\cdot}01 + 0{\cdot}37 \times 8{\cdot}22 = 6{\cdot}75 = w^I(1)$
2	1	$3{\cdot}04 + 0{\cdot}26 \times 6{\cdot}75 + 0{\cdot}40 \times 8{\cdot}22 = 8{\cdot}15$
2	2	$3{\cdot}64 + 0{\cdot}37 \times 6{\cdot}75 + 0{\cdot}23 \times 8{\cdot}22 = 8{\cdot}05$

general effect of descounting is to cause postponement of replacement and since replacement was delayed as long as possible in the undiscounted case this result is not unexpected. In general the optimal policy will vary with the discount factor. An analysis of sensitivity to discount factor is given by Smallwood, *Ops. Res.* 1966, 658–669.

5.6 A RELATED DISCRETE TIME PROCESS

In this section I have drawn on results communicated to me privately by P. J. Schweitzer. For any given semi-Markov process with returns there is a related discrete time Markov process such that the gain rate of the former is the same as the gain of the latter. Consider a semi-Markov process for which the stage return, mean duration of visits and transition probabilities for a general state i are $r(i)$, $t(i)$, $p(i,j)$, $j = 1, \ldots, N$. From equation 5.20 we have for states $i = 1, \ldots, N$

$$v(i) = -gt(i) + r(i) + \sum_{j=1}^{N} p(i, j)v(j) \qquad (5.31)$$

Divide equation 5.31 by $t(i)$ and rearrange the right hand term

$$v(i)/t(i) = -g + (r(i)/t(i)) + \sum_{j \neq i} p(i, j)v(j)/t(i) + p(i, i)v(i)/t(i)$$

Add $v(i)(1/t(i))$ to each side

$$v(i) = -g + (r(i)/t(i)) + \sum_{j \neq i} p(i, j)v(j)/t(i) + (v(i)\,(1 - (1/t(i)) + $$
$$p(i, i)/t(i))) \qquad (5.32)$$

Let

$$p''(i, j) = d_{ij}(1 - 1/t(i)) + p(i, j)/t(i) \qquad (5.33)$$

where $\qquad d_{ij} = 0$ if $j \neq i$ and $d_{ii} = 1$.

Equation 5.32 then becomes

$$v(i) = -g + (r(i)/t(i)) + \sum_{j=1}^{N} p''(i, j)v(j) \qquad (5.34)$$

The sum over j of the $p''(i, j)$ terms is unity. Let the time scale be chosen so that $t(i) \geqslant 1 - p(i, i)$, $i = 1, \ldots, N$. This ensures that $0 \leqslant p''(i, j) \leqslant 1$ for all i and j. Compare equation 5.34 with equation 4.58. We see that equation 5.34 corresponds to a discrete time system with returns $r(i)/t(i)$, transition probabilities $p''(i, j)$ and gain g which is the same as the gain rate of the original semi-Markov system.

Thus a semi-Markov decision problem where for state i and action k the stage return, mean duration of visit and transition probabilities are respectively $r(i, k)$, $t(i, k)$, $p(i, j, k)$ can be converted into a discrete time Markov decision problem with data $r''(i, k)$, $p''(i, j, k)$ where

$$r''(i, k) = r(i, k)/t(i, k)$$
$$p''(i, j, k) = d_{ij}(1 - 1/t(i, k)) + p(i, j, k)/t(i, k) \qquad (5.35)$$
$$t(i, k) \geqslant 1 - p(i, i, k)$$

An advantage of this conversion is that the value iteration algorithm can be applied to the transformed problem. The recurrence relation at iteration stage n is

$$f(n, 1) = \operatorname*{Max}_{k} \left[r''(i, k) + \sum_{j=1}^{N} p''(i, j, k)f(n-1, j) \right] \qquad (5.36)$$

However, if the critical mean visit is much shorter than average the convergence of equation 5.36 will be slow. This is because the modified transition probabilities will be small, except for the $p''(i, i, k)$ which will be close to unity. This tends to mask the difference between policies.

Dynacode: A Dynamic Programming Software Package

Dynacode is a dynamic programming software package developed by the author and written in Fortran. It consists of a main programme and an input/output subroutine. The main programme contains the value interation algorithm of dynamic programming together with a system of conditional statements which enable it to receive and process the data in a way appropriate to the current task. The programme is controlled by a heading card which contains a coded description of the mathematical structure of the current problem.

Dynacode can be applied to problems which involve maximising or minimising additive returns with or without discounting, and which can be deterministic or probabilistic, stationary or non-stationary, finite or infinite horizon and in discrete or continuous time. The scope of the package is thus extremely wide.

A flexible system for handling the input of data and output of results is achieved by the optional use of the input/output subroutine. If the data is in a standard format and the output is required in a standard format the subroutine is not used. In this case a dummy subroutine is compiled. When the input or output is not in standard format a subroutine is written and put in place of the dummy. Some standard subroutines have been developed, e.g., subroutine INVET which is for finite horizon stochastic inventory models. Further details are given by Hastings (1972).

References

BARTLETT, M. S. (1955). *An introduction to Stochastic Processes*. Cambridge University Press.

BECKMANN, M. J. (1968). *Dynamic Programming of Economic Decisions*. Springer Verlag, Berlin.

BELLMAN, R. E. (1957). *Dynamic Programming*. Princeton University Press.

BELLMAN, R. E. (1962). *Adaptive Control Processes*. Oxford University Press.

BELLMAN, R. E. and DREYFUS, S. E. (1962). *Applied Dynamic Programming*. Princeton University Press.

BROWN, B. (1965). 'On the Iterative Method of Dynamic Programming on a Finite Space Discrete Time Markov Process'. *Annals of Mathematical Statistics*, **36**, 1279–1285.

COX, D. R. and MILLER, H. D. (1965). *The Theory of Stochastic Processes*. Methuen.

DENARDO, E. (1970). 'Computing a Bias Optimal Policy in a Discrete Time Markov Decision Problem'. *Operations Research*, **18**, 279–289.

DENARDO, E. and FOX, B. (1968). 'Multichain Markov Renewal Programmes'. *SIAM J Appl. Math*, **16**, 468–487.

FELLER, W. *An Introduction to Probability Theory and its Applications*. Wiley.

HADLEY, G. (1964). *Non-linear and Dynamic Programming*. Addison Wesley.

HAHN, SUSAN G. (1968). 'On the Optimal Cutting of defective sheets'. *Operations Research* **16**, 1100–1114.

HASTINGS, N. A. J. (1968). 'Some notes on dynamic programming and replacement'. *Opl Res. Q.* **19**, 453–464.

HASTINGS, N. A. J. (1969). 'Optimization of Discounted Markov Decision Problems.' *Opl Res. Q.* **20**, 499–500.

HASTINGS, N. A. J. (1970). *Equipment Replacement and the Repair Limit Method. Operational Research in Maintenance;* A. K. S. Jardine, Editor. Manchester University Press/Barnes and Noble Inc.

HASTINGS, N. A. J. (1971). 'Bounds on the gain of a Markov decision process'. *Operations Research*, **19**, 240–244.

HASTINGS, N. A. J. *(1972)*. *Notes on the Dynacode dynamic programming software system.* N. A. J. Hastings, Department of Engineering Production, University of Birmingham, Birmingham, England.

HASTINGS, N. A. J. and MELLO, J. M. C. 'Suboptimality tests in discounted Markov programming' *Mgmt Sec.* (to be published).

HOWARD, R. A. (1960). *Dynamic Programming and Markov Processes.* Wiley.

HOWARD, R. A. (1971). *Dynamic Probabilistic Systems.* Wiley.

JACOBS, O. L. R. (1967). *An Introduction to Dynamic Programming.* Chapman and Hall.

JEWELL, W. S. (1963). 'Markov Renewal Programming I and II, *Operations Research.* **11**, 938–971.

KEMÉNY, J. G. and SNELL, J. L. (1960). *Finite Markov Chains.* Van Nostrand.

MACQUEEN, J. (1966). 'A Modified Dynamic Programming Method for Markovian Decision Processes.' *J. Math. Anal. and Appl.,* **14**, 38–43.

MACQUEEN, J. (1967). 'A Test for Suboptimal Actions in Markovian Decision Problems.' *Operations Research,* **15**, 559–561.

MAKOWER, M. and WILLIAMSON, E. (1967). *Teach Yourself Operational Research.* English Universities Press.

MINE, H. and OSAKI, S. (1970). *Markovian Decision Processes.* Elsevier.

NEMHAUSER, G. L. (1966). *Introduction to Dynamic Programming.* Wiley.

ODONI, A. R. (1969). 'On Finding the Maximal Gain of a Markov Decision Process. *Operations Research,* **17**, 857–860.

PORTEUS, E. L. (1971). 'Some bounds for discounted sequential decision processes'. *Management Science.* **18**, 1 (September), 7–11.

RAIFFA, H. and SCHLAIFER, R. (1961). *Applied Statistical Decision Theory.* Harvard Business School.

ROBERTS, S. M. (1964). *Dynamic Programming in Chemical Engineering and Process Control.* Academic Press.

SCARF, H. E. (1960). *The Optimality of (S, s) Policies in Dynamic Inventory.* Mathematical Methods in the Social Sciences, K. J. Arrow, S. Karlin and P. Suppes, Editors, Stanford University Press.

SCHWEITZER, P. J. (1968). 'Perturbation Theory and Finite Markov Chains' *J. Applied Probability* **5**, 40–413.

SCHWEITZER, P. J. (1971). 'Multiple Policy improvements in undiscounted Markov renewal programming.' *Operations Research* **19**, 784–793.

SHAPIRO, J. F. (1968). 'Tumpike planning horizons for a Markovian decision model.' *Management Science (Theory).* **14**, 5, 292–300.

SMALLWOOD, R. D. (1966). 'Optimal Policy Regions for Markov processes with Discounting'. *Op. Res.* **14**, 658–669.

VEINOTT, A. F. (1966). 'On Finding Optimal Policies in Discrete Dynamic Programming with No Discounting.' *Annals of Mathematical Statistics,* **37**, 1284–1294.

WALD, A. (1950). *Statistical Decision Functions.* Wiley.

WHITE, D. J. (1969). *Dynamic Programming.* Oliver and Boyd/Holden Day.

WHITE, D. J. and NORMAN, J. M. (1969). An Example of Problem Embedding in Deterministic Dynamic Programming. *Op. Res. Q.,* **20**, 469–476.

Index

Network diagram, assortment problem, 58
Networks, 4
  minmax and maxmin problems, 34
Nemhauser, G. L., 69, 82, 168
Nodes, 4, 5
Normalising matrix, 115
Norman, J. M., 21, 168

Odoni, A. R., 136, 168
Optimal path problems, 43
Optimality condition, 29, 33, 35
Optimisation criteria, 125
Ordered set partitioning problem, 57–59
  optimal process, 59
Osaki, S., 124, 128, 153, 168
  problem, 73
Overhaul problem, 73

Partitioning scheme, 115
Periodic problems, *n*-stage, 124
Periodic state, 110, 117
Periodic system, 117
Plan, definition, 3
Policy, 124, 135
  bias optimal, 125, 135–137
  definition, 3
  gain optimal, 125, 128, 135–136
  optimal, policy-value iteration, 161
  optimisation, 126
Policy evaluation operation, 128, 131, 133, 134, 135, 154, 157, 163
Policy improvement routine, 129,131, 133, 134, 135, 154, 157, 163
Policy iteration, 126
Policy-value iteration, 133
Policy-value iteration algorithm, 127–128, 153–159
  flow chart for, 129–130, 155
  gain optimal policy by, 131
  optimal policy by, 161
  solution of fuel pump example, 157
Porteus, E. L., 168
Principle of Optimality, 30
Probabilistic laws, 82
Probability,
  conditional, 83
  transition. *See* Transition probabilities

Probability density of transition, 148, 149
Problem formulation. *See* Formulation
Production, control, 52
Production planning problems, 45–51, 70, 78–80
  optimal process, 51
Progressive problems, 21–24, 97

Raiffa, H., 82, 168
Random variables, 82, 83, 100
Recurrence relation, 10, 11, 13, 16, 22, 46, 54, 58, 61, 66, 70, 86, 88, 89, 91, 94, 97, 98, 119, 136, 138, 139, 165
Recurrent state, 109
Relative bias values, 121–123, 128
Repair limit problem, 105
Repairman problem, 143
Repair truck problem, 144
Replacement problems, 53–56, 74, 105, 144
  optimal process, 56
Return, 46, 150
  additive, 26, 30
  discounted, 159
  Markov processes with, 85–87
  primary, 150
  secondary, 151
  use of term, 4
Return equation, 47
Return function, 61, 66
Return matrix, 56
Risk, situation of, 82
Roberts, S. M., 69, 79, 168
Routing problems, 2–10, 18, 21, 37, 41, 145

s, S policy, 93
Scarf, H. E., 93, 168
Scheduling,
  fishing, 102
  manpower, 77
  overhaul, 73
  production, 70, 78–80
Schlaiffer, R., 82, 168
Schweitzer, P. V., 164, 168
Search problems, 43
Selling a house, 104